PSYCHOLOGY OF EMOTIONS, MOTIVATIONS AND ACTIONS

MODELING HUMAN BEHAVIOR

INDIVIDUALS AND ORGANIZATIONS

PSYCHOLOGY OF EMOTIONS, MOTIVATIONS AND ACTIONS

Additional books in this series can be found on Nova's website under the Series tab.

Additional e-books in this series can be found on Nova's website under the e-book tab.

PSYCHOLOGY OF EMOTIONS, MOTIVATIONS AND ACTIONS

MODELING HUMAN BEHAVIOR

INDIVIDUALS AND ORGANIZATIONS

LUCAS JÓDAR SÁNCHEZ,
ELENA DE LA POZA PLAZA
AND
LUIS ACEDO RODRIGUEZ
EDITORS

nova publishers
New York

Copyright © 2017 by Nova Science Publishers, Inc.

All rights reserved. No part of this book may be reproduced, stored in a retrieval system or transmitted in any form or by any means: electronic, electrostatic, magnetic, tape, mechanical photocopying, recording or otherwise without the written permission of the Publisher.

We have partnered with Copyright Clearance Center to make it easy for you to obtain permissions to reuse content from this publication. Simply navigate to this publication's page on Nova's website and locate the "Get Permission" button below the title description. This button is linked directly to the title's permission page on copyright.com. Alternatively, you can visit copyright.com and search by title, ISBN, or ISSN.

For further questions about using the service on copyright.com, please contact:
Copyright Clearance Center
Phone: +1-(978) 750-8400   Fax: +1-(978) 750-4470   E-mail: info@copyright.com.

NOTICE TO THE READER

The Publisher has taken reasonable care in the preparation of this book, but makes no expressed or implied warranty of any kind and assumes no responsibility for any errors or omissions. No liability is assumed for incidental or consequential damages in connection with or arising out of information contained in this book. The Publisher shall not be liable for any special, consequential, or exemplary damages resulting, in whole or in part, from the readers' use of, or reliance upon, this material. Any parts of this book based on government reports are so indicated and copyright is claimed for those parts to the extent applicable to compilations of such works.

Independent verification should be sought for any data, advice or recommendations contained in this book. In addition, no responsibility is assumed by the publisher for any injury and/or damage to persons or property arising from any methods, products, instructions, ideas or otherwise contained in this publication.

This publication is designed to provide accurate and authoritative information with regard to the subject matter covered herein. It is sold with the clear understanding that the Publisher is not engaged in rendering legal or any other professional services. If legal or any other expert assistance is required, the services of a competent person should be sought. FROM A DECLARATION OF PARTICIPANTS JOINTLY ADOPTED BY A COMMITTEE OF THE AMERICAN BAR ASSOCIATION AND A COMMITTEE OF PUBLISHERS.

Additional color graphics may be available in the e-book version of this book.

Library of Congress Cataloging-in-Publication Data

ISBN: 978-1-53610-197-3

Published by Nova Science Publishers, Inc. † New York

CONTENTS

Preface vii

Chapter 1 Why the Spanish Public University Model Is Wrong: Causes and Recommendations for Improvement 1
Lucas Jódar Sánchez

Chapter 2 A Case Study of Directional Communities in a Directed Graph: The Accessing Procedure to the Spanish Public University System 9
A. Hervás, P. P. Soriano-Jiménez, A. Jiménez, J. Peinado, R. Capilla and J. M. Montañana

Chapter 3 Validation of Incode Framework for Assessment of Innovation Competency of Higher Education Students: A Multidimensional Technique for Affinity Diagram to Detect the Most Relevant Behaviours and Skills 25
Mónica Martínez-Gómez, Manuel Marí-Benlloch and Juan A. Marin-Garcia

Chapter 4 Evaluation of M-Learning among Students According to Their Behaviour with Apps 37
Laura Briz-Ponce, Anabela Pereira, Juan Antonio Juanes-Méndez and Francisco José García-Peñalvo

Chapter 5 Assessing University Stakeholders Attributes: A Participative Leadership Approach 49
Martín A. Pantoja, María del P. Rodríguez and Andrés Carrión

Chapter 6 Intervention Programme for Pharmacy Office Preventing Metabolic Syndrome: Improving the Population's Quality of Life by Modelling Its Behavior 57
María del Mar Meliá Santarrufina and Fernando Figueroa

Chapter 7 Actors and Factors Involved in Health Technology Diffusion and Adoption: Economic, Social and Technological Determinants 65
María Caballer-Tarazona and Cristina Pardo-García

Chapter 8	Assessing the Operation of and User Satisfaction with the Electronic Prescribing System in the Valencian Community (Spain) *Isabel Barrachina, Elena de la Poza Plaza, Beatriz Pedrós and David Vivas*	85
Chapter 9	Modelling Human Behaviours by Shaping Organizational Culture *Mateusz Molasy*	99
Chapter 10	Robbery Attractiveness among Urban Areas: A Computational Modelling Approach *R. Cervelló-Royo, E. Conca-Casanova, J.-C. Cortés and Rafael-J. Villanueva*	113
Chapter 11	The Peak Work of the Patriarch Ribera in the Counter-Reformation: The Royal Seminary-School of Corpus Christi of Valencia (Spain) *Carlos Lerma, Ángeles Mas, Enrique Gil and Jose Vercher*	125
Chapter 12	Modeling of Human Capital and Impact on EU Regional Competitiveness *Lenka Fojtíková, Michaela Staníčková and Lukáš Melecký*	133
Chapter 13	The Cox-Ingersoll-Ross Interest Rate Model Revisited: Some Motivations and Applications *R.-V. Arévalo, J.-C. Cortés and R.-J. Villanueva*	165
Chapter 14	Consumers' Multi-Homing or Multiple Demand *Cristina Pardo-García and María Caballer-Tarazona*	177
Chapter 15	Modelling Learning under Random Conditions with Cellular Automata *L. Acedo*	185
Chapter 16	Capturing the Subjacent Risk of death from a Population: The Wavelet Approximation *I. Baeza-Sampere and F. G. Morillas-Jurado*	195
Chapter 17	Trajectories Similarity: A Proposal and Some Problems *Francisco Javier Moreno, Santiago Román Fernández and Vania Bogorny*	211
Chapter 18	A Tensor Model for Automated Production Lines Based on Probabilistic Sub-Cycle Times *E. Garcia and N. Montes*	221
Chapter 19	Building Lifetime Heterosexual Partner Networks *L. Acedo, R. Martí, F. Palmi, V. Sancehz-Alonso, F. J. Santonja, Rafael-J. Villanueva and J. Villanueva-Oller*	235
About the Editors		251
Index		253

PREFACE

This book is devoted to model human behaviour from the individual and collective point of view. Both approaches are connected because as humans, we imitate our peers in an attempt to socialize and integrate ourselves at organizations but also we compare ourselves with others as part of an internal and external process of competition.

The first five chapters are related to educational issues. In the first chapter Jódar analyzes the government of the Spanish public university system. Causes of wrong behaviour, risks and recommendations for improvement are presented. In chapter 2, Hervás et al. propose a model for students to access Spanish public universities allowing a measure of the system performance. In Chapter 3 authors Martínez-Gómez et al. develop a qualitative model to identify innovative skills of higher education students. The authors Briz-Ponce et al., consider in chapter 4 the convenience and possibilities of introducing mobile techniques for learning improvement at higher education level. A leadership model for University stakeholders of several Colombian public universities of the city of Manizales is proposed in chapter 5.

Chapters 6, 7 and 8 are focused on healthcare models. In particular chapter 6 by Meliá and Figueroa presents an intervention programme to implement healthy habits in human behavior from a pharmacy office. Preventing and controlling metabolic syndrome improve population's quality of life of patients.

In chapter 7, Caballer and Pardo identify and classify the main factors involved in the process of adoption and diffusion of technology, affecting healthcare quality and performance. In chapter 8, Barrachina et al. analyze patients' satisfaction with implementing system of electronic prescription. They identify how the satisfaction of patients with chronic diseases with the system relies on having to go to their medical centre less frequently and the time spent with one's doctor not being cut.

In chapter 9, by combining the achievement of both psychological, sociological and anthropologic factors, Molasy proposes a human behaviour model by shaping organizational culture.

In chapter 10, Cervelló et al. construct a behavioural model to predict the attractiveness for burglars of different city areas.

In chapter 11, Lerma et al. links the architecture style with cultural behaviour during the Renaissance period at the city of Valencia.

Following chapters 12, 13 and 14 are of economic nature. In concrete Fojtikova et al., analyze the role and the significance of human capital on competitiveness of several EU regional economies.

In chapter 13, Arévalo et al. propose a revisted Cox-Ingersoll-Ross (CIR) interest model applied to predict the Euribor interest rate from a real sample including some measures of goodness of fit. On contrast, Pardo and Caballer in chapter 14 analyze the impact of multiple purchase behaviour by consumers on firms.

In Chapter 15 Acedo studies the learning process from the point of view of neural networks obtaining some limits about the ability to learn and generalize from imperfect data sources. Baeza et al. build in chapter 16 a wavelet model to fit death rates in a population, which can be of interest to biometricians.

In chapter 17, Moreno et al. proposes a model for tourist behaviour determining the similarity with regard to visited places.

García and Montes, presents in chapter 18 a tensor model for automated production lines based on probabilistic cycle times. Validation is performed using data from a car factory located at Almussafes, (Valencia, Spain). Finally, in chapter 19 a model for a sexual network is developed from realistic statistical data with the long term objective of applying it to the control and prevention of sexually transmitted diseases.

Chapter 1

WHY THE SPANISH PUBLIC UNIVERSITY MODEL IS WRONG: CAUSES AND RECOMMENDATIONS FOR IMPROVEMENT

Lucas Jódar
Instituto de Matemática Multidisciplinar, IMM
Universitat Politècnica de València,
Ciudad Politècnica de la Innovación, Valencia, Spain

ABSTRACT

This chapter analyses the faults of both the governing model and the management of Spanish public universities, and the risks of its continuous deterioration. It offers possible recommendations for improvement.

Keywords: public university system, wrong behaviour, causes, recommendations

1. INTRODUCTION

After 38 years of research and teaching experience, having been involved at all academic, department management and university research institute levels, and having visited universities in other countries, I have identified several severe faults in the Spanish public university system which I would like to share and diffuse.

Encouragement is constructive, but with little hope because apart from other serious problems, as we all know, in Spain we distinguish ourselves for political parties not being capable of reaching an agreement on a stable law for education. It is a matter in which loss of sovereignty would surely imply improvement. If the European contagion influenced us, things would no doubt improve. Whether we like it or not, losing sovereignty does not always have to be negative.

Until this change in the law comes about, this chapter analyses several serious problems that I believe should be dealt with:

i. Mechanism by which rectors are elected
ii. Ageing teachers
iii. Science policy. Selecting teachers and students

This chapter is arranged as follows: Section 2 analyses the possible causes of today's situation by paying special attention to details and to the seriousness of the aforementioned problems. Section 3 offers some recommendations for improvements.

2. CAUSES, EFFECTS AND SERIOUSNESS OF PROBLEMS

Although as humans we boast about rationality, it is merely an individual and group illusion, and frequently implies emotions that are not always generous or altruistic that dominate our acts. Revanchism, a party's interests, cancelling former governing parties' acts ("Adamism")[1] are, unfortunately, all too frequent. Wars are a good example, but such behaviours are, to a lesser extent, frequent in many alternations in power.

So often in history, in both Spain and beyond, one era of excesses is, in some sense, succeeded by another. Here excesses are understood in the converse ideological sense which, evidently, if not stopped, will once again make the situation worse, if indeed they actually improve anything.

Balancing this imaginary pendulum of history is never accomplished because asking politicians to reach agreements about important matters is difficult in general, and nowadays represents a Utopia in Spain. It is a matter for mature democratic societies which, unfortunately, is not Spain's case.

After almost 40 years of dictatorship in Spain, since 1978 and after some years of natural turmoil, the arrival of the new Spanish Constitution sorted many matters out, but quite a few other major items still need solving. With lack of democracy during the dictatorial period, it was succeeded (and its perverse effects still live on) by over-democratisation, which inundated scenarios where the democratic criterion is not only not convenient, but is also perverse.

So the question is, what must, and must not, be democratically elected? It is no easy question to answer. Fortunately in our western culture, some sections of freedom remain, which we, as individuals, can afford.

For instance, what vocation to choose, or where and with whom do we wish to share our life, is not democratically chosen by any electoral body. Which is just as well because no-one would ever consider making all decisions in a family democratically because not all family members have the same level of knowledge and capacity. Is a 10-year-old's opinion worth the same as his/her parent's view? Perhaps it is for some things, but certainly not in most decisions or for the most important issues.

---

[1] Translator's note: Adamism, from the Spanish "adanismo," meaning: Habit of initiating any activity as if no-one had ever done it before.

So no-one acuses me of not having an opinion, I must say that in institutions which people enter freely, the opinions of those who were already there may prevail over those of newcomers. This initial situation must be respected, and then people can convince others and change it once inside.

When people become University Community members, they do so willingly and the institution already existed. The opinions of those who were already there may prevail over those of newcomers. This initial situation must be respected, and no-one can assume an "equality" right in the precise moment they form part of it.

The University must not be governed democratically because it must seek excellence rather than teachers' equality. The University must motivate performance and acknowledge it, and cannot, therefore, be on the same footing, but should favour performance inequality because that is where excellence can emerge from. Favouring equality leads to mediocrity.

The human condition is such that we constantly compare ourselves with our peers, who we mimic, and from whom habits and customs are passed on to us [Girard, Raafa, Christakis]. If when you compare yourself to others you do not see those who stand out, it is not worth making the effort. If those who stand out are not acknowledged, no-one will stand out. So no effort or improvement will be encouraged, and excellence will never occur.

The public University is not a company, but has many similarities to one, or at least should have them. A good company not only seeks profits, but customer satisfaction because it is the best way to achieve customer fidelity and to, therefore, ensure that the company continues. Companies that survive and last are characterised by prioritising customer satisfaction. Those that only seek to make profits disappear [P. Olmreod, Why most things fail].

The company that does not work ends up closing and disappearing, but this not the case of the Spanish public University. More importantly, a high percentage of public University employees do not work well, and practically no-one leaves.

How can a Spanish public University rector implement measures that encourage making efforts and excellence if (s)he is to be elected by a generally mediocre majority?

The thing is, in order to be chosen, populist measures are often taken during electoral periods so rectors can be re-elected. These measures may have a very negative effect. This is known as populism of rectors, which is strikingly similar to populism of professional politicians.

By way of example, in order to be voted by many non-PhD teachers, their teaching load is lightened so they can do their doctoral studies. The teaching that they do not undertake is covered by contracting unnecessary teachers according to as many lax criteria as necessary so that everything is legal and appearances are maintained.

This incurs additional and unnecessary expenditure on personnel, and on badly selected personnel. Another example, backed by directors, is to approve bonuses to administration personnel for their usual work, which normally forms part of their competences. If you would like more examples, financing students, usually their representatives, and any budget, premises or building, are to name but a few.

When a crises starts and the central government makes cuts, rectors complain that they cannot contract more people.

The result is that no-one is contracted for many years. The few that are contracted, many are mediocre because choice of personnel was not based on excellence, but on voters' social content.

This situation is so serious that it is quite likely that nobody young is contracted for a decade, no matter how brilliant they are. So the existing personnel ages and quality lowers.

This situation is very serious because all trade unions of all kinds, given the aberrant class of full-time members who are excluded from their usual work to simply and totally work as politicians so that civil servants' privileges continue, dare to think that the public University could be politicised if the rector is not democratically elected.

What happens nowadays is that the whole University is politicised to maintain, starting with the rector, directors of centres, heads of departments, and trade union representatives. One year before rector elections, an army of interested parties is mobilised to place his/her representatives.

Can excellence exist in such a situation? Of course not, and it will never exist if things go on this way. The few cases of excellence that sprout survive even when they go against the tide. The amount of time continuously wasted in committees of all kinds to maintain hegemonic positions is endless.

The bureaucratisation of almost everything, and maintaining all kinds of privileges, are the most important issues, no matter where. If all this was not enough, as those who work in power positions never rest, they make the best of any kind of advantages because it is they who actually decide them. Lightening teaching workloads is not based especially on research excellence, but on management posts. Full-time trade union members, and there are hundreds of them, do not do their normal work because they are exclusively and politically active.

Although working hours and objectives in a private company are controlled by the company owner because (s)he takes care of them, practically no-one controls if work is being done or if working hours are fulfilled in the public University, apart from classes because students would complain if teachers were absent. Nobody complains that most of the administration and services personnel do not work all their working hours because it is gets you nowhere.

Working hours are not controlled in many universities. Starting work half an hour late is an everyday occurrence for administration and services personnel. When they know what time their political bosses arrive, they start work a bit earlier, and that's the end to it. This implies a slightly more than 5% waste of personnel expenditure.

Trade unions are not there to ensure that working hours are worked, but to protect those who do not fulfil their duties. Absenteeism from work is a permanent widespread practice.

All these public University dysfunctions become even worse at Secondary Education levels, where the personnel is not assessed, and where the army of trade union representatives is concerned about things continuing this way to maintain the "everything goes" privilege and a holiday period that nobody else has. In July and August, Spanish Secondary schools perform hardly any activities.

I do not exaggerate when I state that Secondary Education teachers have 1 month more holidays that all other citizens. This represents more than 7% of wasted public expenditure.

Every year students are less prepared to start University. There are many reasons for this, but one is due to inadequate teaching activity, whose main concern is certainly not students' optimum educational preparation. If no-one assesses them, no-one has any reason to demonstrate their deficiencies.

Unproductive Public Policies and Possible Alternatives

The public University may have some virtues, but it would not be fair to compare it with the private University because the former is financed mainly by public funds, while the latter is not.

As other authors have maintained [4, 5], the State may have usefully corrected market failures, or maintained a level of innovation by encouraging basic research, but this certainly does not mean that money has been well spent.

The above-mentioned defects of the Spanish Public University are due mainly to faulty education and science policies. I do not believe that it is because Spaniards are less capable than the Germans, British, French or North Americans. It is rather due to the fact that Spanish institutions, and education and science policies, are worse.

It is true that public expenditure in Spain is considerably lower than it is in leading countries. However, it is not merely a matter of spending, but spending well. In the previous section, we showed that more than 5% of personnel expenditure is wasted in public Universities, and that does not include the money wasted on full-time trade union members and absenteeism from work.

We also mentioned that Secondary Education wastes more than 7% by taking into account only the hours not worked in July. No doubt trade union representatives, who merely defend public privileges, state that teachers are training and on courses, etc., in July. That is all well and good, but this does not mean that they have to be absent from work every day.

Since about 1980, the two main political parties in Spain, PSOE (the Socialist Party) and PP (the Popular Party), have monopolised the State by flooding public institutions with others who think like them [6]. There has been plenty of stability, no doubt too much of it. For 35 years, both PSOE and PP have not been capable of reaching a consensus about laws of education and research. And so it is that we now face a grave situation at nearly all levels of education.

For a long time now I have wondered, why is that when the economic situation in our leading neighbouring countries is bad, in Spain things are much worse?

I believe that our institutions, laws and culture make us drowsy, they protect us against risks, they do not tell us the truth, and they make us incapable of effort and innovation. The Spanish wake up every morning convinced that the State has to provide them with health, education, a home, and miniumum sustenance, and all free of charge. But where do all these guaranteed resources for so many people come from?

For a long time politicians have not told the truth, probably because they thought that we would not vote them if they did. This attitude of Spanish leaders, with this overprotective culture, does not encourage citizens to offer their very best. When things do not go smoothly and we face difficulties, we do not know what to do. We are drowsy.

The consequences are worse still because they do not favour ethical behaviour among civil servants. It is not unusual to come across colleagues who do not wonder, what do I owe the State so that it guarantees my job? Quite the opposite occurs in fact as they wonder what they can obtain from the State. The main objective is not service, and its objective does not even have to coincide with that of everyone else. The State only exists to provide me and not to expect something of me.

Such a perverse principle underlies the laxness of respecting what is public, and is the starting point of being permissive with corruption. Arriving at work later on a regular basis is also corruption.

Another perverse public principle is there is no retroactivity from awarding bonuses. This principle does not favour maintaining productivity, but instead, the principle of minimum effort instead in many cases. It is an almost general rule of thumb that productivity drops after some progress has been made.

The correct principle should be that, apart from one's basic salary, which should be guaranteed, productivity bonuses should remain if productivity continues, but should lower if productivity does not continue. Civil servants, paid by all citizens' taxes, must be an example, and not otherwise as they exploit privileges that private workers have no access to.

The origin of this mine of privileges is none other than the populism repeated by the two main political parties that have governed the country since 1980. Naturally, they are not the only ones. The thousands of people with parliamentary immunity are another example of embarassing privileges.

Although the two main political parties are chiefly responsible for all this, which have governed Spain alternately since 1980, I believe that left-wing governments are more responsible because they are all for public expenditure being more efficient, which it certainly cannot be with the above-cited privileges.

Political parties must know human beings, must constantly encourage productivity, assess well and demand performance. Maintaining public services like education, a health system and pensions costs a fortune, and every public euro must be efficiently managed.

There is now much talk about political regeneration, but no-one talks about redirecting the public service in the right direction. It is not easy to imagine that this could happen in the short term because the party that proposes it would lose votes. However, a possible criterion would be to agree on a long-term change in public administrations, say in 10 years times, so there would be fewer people affected and they would be warned.

There are too many politicians, but judges and tax inspectors are lacking. Civil servants have too many privileges, but there is a shortage of public day nurseries. It is necessary to raise birth rates and to look after the elderly, but there are too many universities.

Another reason for inefficiency is subsidising students who are mediocre or lack motivation. University studies are not an asset for our survival, so citizens must not pay the cost of university education of students whose families can afford university fees. A brilliant student whose family cannot afford these fees must receive a grant, but only those who excel, and not mediocre students because they do not often come to class.

Left-wing parties and their talk of equality have considerably damaged, and continue to do so, the whole country. The same is true of conservatives as they have not defended what is right because it is complex, or have simply looked the other way. Equal opportunities can never be achieved in this way because, if there is a superpopulation with degrees, degrees are not acknowledged and deteriorate. Spain does not take educational issues seriously because degrees are worthless and, among many other things, there is a high level of academic failure.

Why is academic failure so widespread and at all levels of education?

The selection of teaching staff must be much more demanding, and the assessments made of them must be permanent and demanding, where public service is a priority rather than using the State for their own benefit. Secondary education teachers who are not assessed can become civil servants too easily. Their income is much better than it is at universities, and I do not mean it should be worse, but much less is expected of them.

Most teachers want no problems, mainly because they do not have the authority and prestige they should have. If any of them take the initiative, a whole army of well-to-do people can make their life at their workplace extremely difficult. The principle of the State acting to serve them has taken root.

Apart from suitable contents, about which there is a great deal to talk, what has been agreed on is simply not taught. Teachers keep away from problems, and the best efforts principle is lacking in teachers and students alike. Rather than teaching what they should, institutes simply prepare students to sit university entrance exams in order to obtain good results from them. And disaster is certain as these tests are poorly designed.

If the best efforts culture is not alive at home, at school, indeed not anywhere, what can we expect?

The consequences are not only technical in nature, but are more profound and respond to a culture that does not encourage making efforts, taking risks, innovating, respecting what is public or ethical behaviour. These shortages cover up very serious problems, such as violence and school bullying, which is a matter that certainly needs to be dealt with separately.

REFERENCES

[1] R. Girard, Mimesis and Theory: Essays on Literature and Criticism, 1953-2005, Stanford Univ. Press, 2008.
[2] R.M. Raafat, N. Chater, and C. Frith, Herding in humans, Trends in Cognitive Sciences, vol. 13, no.10 (2009), pp.420-428.
[3] N.A. Christakis and J.H. Fowler, Connected: The Surprising Power in Our Social Networks and How They Shape Our Lives, Bac Bey Books, Little Brown and Company, USA, 2009.
[4] M. Mazzucato, The Entrepeneurial State, Anthem Press, 2014.
[5] J.M. Martín Carretero, España 2030: Gobernar el Futuro, E. Deusto, 2016.
[6] A. Nieto, El desgobierno de lo público, E. Ariel, 2012.

In: Modeling Human Behavior: Individuals and Organizations ISBN: 978-1-53610-197-3
Editors: L. Jódar Sánchez, E. de la Poza Plaza et al. © 2017 Nova Science Publishers, Inc.

Chapter 2

A Case Study of Directional Communities in a Directed Graph: The Accessing Procedure to the Spanish Public University System

A. Hervás[1,3,*], *P. P. Soriano-Jiménez*[3,†], *A. Jiménez*[3,‡], *J. Peinado*[2,3,§]
R. Capilla[3,¶] *and J. M. Montañana*[4,‖]

[1]Instituto de Matemática Multidisciplinar (IMM),
Valencia, Spain
[2]Instituto de Instrumentacion para la Imagen Molecular,
Valencia, Spain
[3]Universitat Politecnica de Valencia, Valencia, Spain
[4]Department of Computer Science,
University of York, Heslington, York, UK

Abstract

The procedure for the access to the Spanish Public University System, S.U.P.E., is a complex process in which students request for placing in several degrees, and the system assigns each applicant a degree and a university according to the criteria established by law. During the process, student traffic between different degrees is produced. Knowing the structure of the traffic between degrees in the access process and characterising the degrees based on this traffic, can be a useful tool for universities in order to decide regarding the development and the modification of distribution plans of the new offers. This paper proposes firstly an algorithm to represent in a directional graph the traffic of students in the process of access, and secondly we have proposed an algorithm that allows us to group the vertices of the graph in directed communities and study their properties.

---

[*]E-mail address: ahervas@imm.upv.es
[†]E-mail adress: psoriano@upv.es
[‡]E-mail adress: anjimgar@doctor.upv.es
[§]E-mail adress: jpeinado@dsic.upv.es
[¶]E-mail adress: rcapilla@eln.upv.es
[‖]E-mail adress: jmmontanana@gmail.com

Keywords: higher education management, social models, graphs and networks applications, clustering, cluster analysis, complex networks

AMS Subject Classification: 97D60, 91C20, 97B40, 90B18, 62H30

1. Introduction

The access procedure to the Spanish Public University System (S.U.P.E.) is a complex process that differs from both the Spanish private system, and other university systems.

Every year, each university offers a number of places for each one of their degrees. Those are the places available for the new first-year students.

Once finished High School, students must pass an exam, and then, according to a polynomial formula combining the results of this examination, the marks obtained in High School and some subjects related with the grades each student wishes to apply, they must select in an ordered way their degree choices. Therefore, student X applies for Degree A in first position, degree F in second place, another degree D in third place, and successively.

The number of students participating in the process depend on regions. For instance, in Catalonia over 50.000 per year, in the Valencian Community over 25.000, and only in Madrid over 38.000 students.

The "system" assigns each applicant a specific degree and a university based on the criteria established by the laws at every regional government. However, in case of not being assigned to any degree, the student will be included on a waiting list.

Thus, student X is assigned a place in the grade A, or grade B, or grade C. If the choice has not been adequate, he will not obtain a place in any degree. If he gets a place in his choice number n, he stays in an ordered waiting list in each of his previous $n-1$ choices, so if there are drop outs, the list scrolls and causes movements in other waiting lists.

The matter is that those students obtaining the best grades, get their first election. On the other hand, the rest of students get their best choice where places are available.

Consequently, lists movements take place between those students who have requested a degree and had been allocated to a less desired degree.

The movement of positions occurs in few days, and it can involve a large number of students. In some cases, there are rotary movements that require several iterations until the system is finally stabilized.

Moreover, there are little movements in the list in high and low-demanded degrees, while the traffic in degrees with middle demand is very intense. Consequently, at the end of the process there are some degrees that have been chosen as a first option by most of the students. On the other hand, there are other degrees with few students who have chosen it as their first option. The excess of demand over the offer of places in a degree is distributed among the degrees with lower demand.

Knowing the structure of how the first-choice demand works and the traffic between degrees and universities, may become into a very useful tool in the design of the offer of places. So it is when designing or restructuring the map of degrees of a regional government, understanding other aspects related to the functioning of some degrees as the performance, dropout, transfers, absenteeism, etc.

This paper presents a process that allows us to model the S.U.P.E. system access, in order to identify some properties of the degrees and, consequently, to analyze the abstract performance of the system.

2. Preliminaries

The first approach to the problem leads us to a structure in which traffic between the elements can trigger a multi directional traffic, and where the flow may be extended between various elements influencing each other.

This leads us to consider the use of graphs and networks as a tool to approach the issue.

Graph theory has been used to establish and solve many problems. From its beginnings in the eighteenth century, concerning the problem of the bridges of Königsberg -studied by Euler-, the conjecture of the four colors, the problem of "maximum flow-minimum cut", the Chinese postman problem, the transportation problem, the assignment problem or combinatorial problems among others. [11], [9] and [1].

In recent years, due to the development of the WWW, the growth of social networks and the explosion of Big Data, the study of complex networks has led us to an interesting and fruitful line of work for scientists. It allowed applications in many fields, from the area of biomedicine to the social sciences: genetics, study of epidemics, coauthoring publications, social relations, etc.

These networks are characterized by the analysis of the degree of the vertices and the existence of paths or cycles -shorter or longer- between pairs of vertices.

The study of the structure of these networks helps us to understand how they work and allows us to create new models, hitherto not posed by the classical theory of graphs. Adding the dynamic behavior of these structures, we have a complex collection of interesting problems.

An excellent and comprehensive review of the structure and dynamics of complex networks can be found in [3].

It acquires special importance the study of those elements that are closely related. It is created a cluster of elements in the structure that are highly connected with each other and few connected with the rest. We will call these groups communities, and they involve a structure within the network.

The finding of communities on a network allows us to approach the knowledge of the structure and its behavior. The design of algorithms will allow us to obtain communities in a network, and it will become a key point of work to be developed. In this regard, we must highlight the work of [8], [14], [15] and especially [7], an excellent and complete review of the state of the art in modularity, clustering and its applications.

3. Building the Graph

The first approach to the problem leads us to a structure in which traffic between the elements can trigger a multi directional traffic, and where the flow may be extended between various elements influencing each other.

This leads us to consider the use of graphs and networks as a tool to approach the issue.

Graph theory has been used to establish and solve many problems. From its beginnings in the eighteenth century, concerning the problem of the bridges of Königsberg -studied by Euler-, the conjecture of the four colors, the problem of "maximum flow-minimum cut", the Chinese postman problem, the transportation problem, the assignment problem or combinatorial problems among others. [11], [9] and [1].

In recent years, due to the development of the WWW, the growth of social networks and the explosion of Big Data, the study of complex networks has led us to an interesting and fruitful line of work for scientists. It allowed applications in many fields, from the area of biomedicine to the social sciences: genetics, study of epidemics, coauthoring publications, social relations, etc.

These networks are characterized by the analysis of the degree of the vertices and the existence of paths or cycles -shorter or longer- between pairs of vertices.

The study of the structure of these networks helps us to understand how they work and allows us to create new models, hitherto not posed by the classical theory of graphs. Adding the dynamic behavior of these structures, we have a complex collection of interesting problems.

An excellent and comprehensive review of the structure and dynamics of complex networks can be found in [3].

It acquires special importance the study of those elements that are closely related. It is created a cluster of elements in the structure that are highly connected with each other and few connected with the rest. We will call these groups communities, and they involve a structure within the network.

The finding of communities on a network allows us to approach the knowledge of the structure and its behavior. The design of algorithms will allow us to obtain communities in a network, and it will become a key point of work to be developed. In this regard, we must highlight the work of [8], [14], [15] and especially [7], an excellent and complete review of the state of the art in modularity, clustering and its applications.

Algorithm 3.1 (Construction of the Graph). This algorithm constructs a graph representing the degree of SUPE that are given in an autonomous community and the students applications in the process of pre-registration to the University in a given year.

1. Let be $V = \{v_i, i \in I\}$ Set of degrees offered by the Regional Governement[1]

2. Let be $S =$ Set of applications from students in the process of pre-registration of a given year.

3. We define the graph $G = (V, E)$. We start from $E = \emptyset$.

4. We calculate the vector $f = (f_1, f_2, f_3, ..., f_n)$ where f_j is the frequency of applications accepted in the position j.

5. We choose $i*$ like the one value for which $\sum_{j=1}^{i*} f_i > 0.9$ and $\sum_{j=1}^{i*-1} f_i < 0.9$

6. Obtaining edges

---

[1] Each vertex and edge are encoded by adding the necessary attributes for further identification and processing.

(a) A student i choose the options $(v_{i1}, v_{i2}, ..., v_{ii*})$,

(b) If the edge $(v_{kj}, v_{kj+1}) \in E$, then increase their weight $P_{(v_{kj}, v_{kj+1})} = P_{(v_{kj}, v_{kj+1})} + f_{j+1}$

(c) If the edge (v_{kj}, v_{kj+1}) is not defined, then we add the edge to the graph and its weight should become $P_{(v_{kj}, v_{kj+1})} = P_{(v_{kj}, v_{kj+1})} + f_{j+1} = f_{j+1}$

7. Once obtained the graph, we will remove the edges that do not provide relevant information.

8. We calculate the maximum threshold Modularity

9. For every vertex v_k y v_l,

    (a) We remove the edges (v_k, v_l) whose weight is lower than the threshold, unless the edge is the one with maximum in-weight or out-weight in v_k.

    (b) If there were more of an edge, and below the threshold, we would keep the most weight.

10. Thus, it was obtained the graph $G = \{V, E\}$

In Figure 1 we can see the graph built using Algorithm 1. The graph obtained is a directed graph, where the size of the vertices indicates the number of vacancies offered to students; the edges indicate the direction of flow of students; and the weight of the edges indicates the value of the flow according to the criteria discussed above.

4. Graph Structure

The resulting graph is unconnected. We can observe a certain dispersion, although large aggregations appear and other result quite small, with few vertices. The results are similar in other regions.

Once obtained the graph, the first natural step is to study their connected components.

For the graph in Figure 1, we obtain 131 strongly connected components; and its 12 weakly related components. These components are represented in Figure 2 by surrounding them with a line, by filling in the area of color.

In our case, it must be highlighted that the graph represents the traffic of students between grades taught in a region. It is interesting to perform a short analysis by grouping the degrees by areas. Thus, if we consider the subgraph formed by engineering degrees and the vertices connected with them, see Figure 3, we observe the existence of large groups, with academic sense, given the priority to access a title regardless of the University, or others of similar characteristics. There are also small groups, which can be either geographical, students who prefer to study in a particular environment or closely academically related degrees. See Figure 3, for a group of 8 ITC degrees just 30 minutes away, and Figure 4, for the same degree in two different universities and a close degree in the same area. This occurs in a similar way for the different areas of degrees.

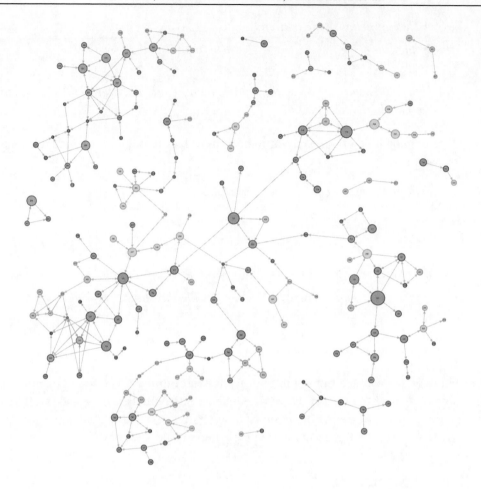

Figure 1. The graph for a system with 250 degrees and 25,000 students.

Moreover, we observe that larger vertices have higher degrees than the smallers. This makes sense, since the higher number of places offered, the more possibilities they have of second election and, therefore, of traffic distribution.

Weakly connected components or very small subgraphs supply us, corresponding to a defined geographical or academic environment, or very large subgraphs that do not provide students traffic information.

The strongly connected components provide us many subgraphs that are strongly connected internally. However, they offer few information about the students traffic since if the traffic between two vertices only occurs in one direction, even if it is very intense, the strong connection separates those vertices and we lose relevant information about the problem (see the subgraph of Figure 4).

It can also be highlighted that there are vertices with only one input degree, and other vertices that have a small degree, one or two, but that serve as union between groups of respectable size. This makes us think that we should study in greater detail the characteristics of the graph, but we need new tools to do so. See vertices 203 and 207 from Figure 5, on the upper left corner of the graph, corresponding to engineering degrees.

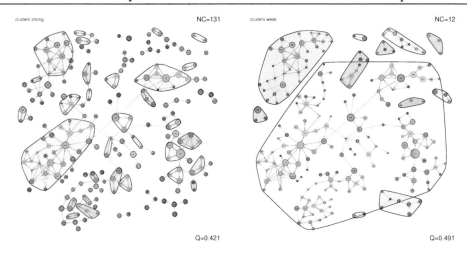

Figure 2. Strong and weakly connected component of the graph of Figure 1.

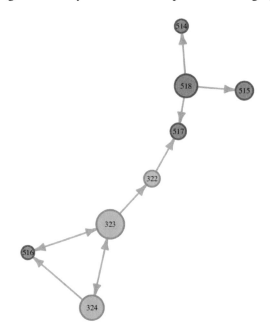

Figure 3. Connected component with geographical relationship.

Thus, we study the graph over the point of view of communities. This concept is related to the high density of connections in the graph, [7], [8], [14].

In a graph, a community composes a set of vertices that are highly interrelated, meaning that there are many edges between them. In contrast, there are few edges that connect the community vertices with the rest of the graph.

In other words, there is a high density of connections within each community and a low density of connections between communities. The reason for using this technique is given by the fact that: *"Community structure methods normally assume that the network of interest divides naturally into subgroups and the experimenter's job is to find those groups.*

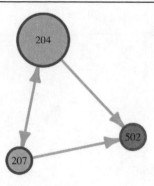

Figure 4. Connected component with academic relationship.

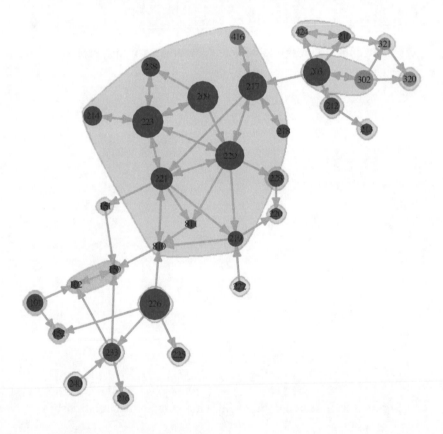

Figure 5. Strong connected components of the graph corresponding to the area of Engineering.

The number and size of the groups are thus determined by the network itself and not by the experimenter.", [14].

Given a network, a good division in communities is the one that gets a large number of edges within the communities, against a small number of inter-community edges (the incidents vertices of these edges belong to different communities).

It is considered a good division the one that gets communities to have a large number of edges within the small communities facing a number of inter-communal edges (incidents vertices these edges belong to different communities).

Modularity is a property that indicates how good this division is. Modularity is a function that evaluates the goodness of partitions of a graph into clusters.[7], [15]. Takes value between 0 and 1, and according to [15], in practice, networks that have a strong community structure have a modularity between $0,3$ and $0,7$. Higher values are strange.

There are several algorithms that allow us to get the communities on a graph, and show us modularity. Nonetheless, they offer different results depending on the criteria for group vertices used. Most of them only apply to undirected graphs.[7], [8], [5], [21] and [16].

Walktrap, proposed by [19], works by joining communities through random walks.

In label propagation, proposed by [21], the algorithm assigns labels to the vertices that are updated at each iteration. Although it provides good computational results, it does not offer a unique solution.

The edge betweenees, proposed by [15], begins with only one community, and divides it, until obtaining n communities.

The fast greedy, by [16], improves the computational results of edge betweenees. Assumes that each vertex is a community, grouping them at each, ending with n communities

We apply all these algorithms to our graph. Despite the community obtained looking similar, the real results are very different. See Figure 6.

In Figure 6, for each graph, it is shown in the upper left corner the algorithm used,in the upper right the number of communities obtained, and its modularity in the lower right corner . Although the results obtained give excellent modularity values, above 0.8, and a similar number of communities, the results are substantially different and do not add an additional value to what discussed above. This might be probably due to these algorithms are designed for undirected graphs, although they have been successfully used in some cases for directed graphs.

This is the reason why we discard the use of these algorithms, since our goal is to establish a method to obtain the communities so that we can detect those vertices on which to act on them alter the system. If we could get it, the next step would be to develop procedures to analyze changes in the system, the graph, as of minor changes to the appropriate vertices.

5. Proposed Algorithm

So, we propose an algorithm that finds communities in which all vertices can reach the same vertex, or be reached. Considering the directionality of the graph, it leads necessarily to study the two possibilities: communities vertices that can be reached from a given one, and communities of vertices from which it has reached the vertex considered.

Algorithm 5.1 (Construction of Directional Communities). With this algorithm we will obtain the community sets of vertices that allow us to generate subgraphs making up the communities.

1. *From the graph obtained in the previous algorithm:*

2. *Applying a search algorithm we obtain the matrices of accessibility, R, and access, Q, of the graph.*

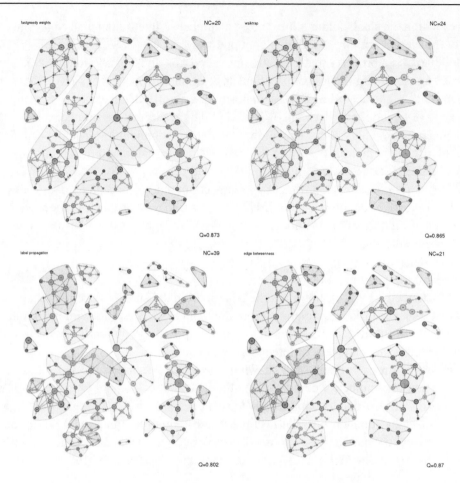

Figure 6. Community graph under 4 different algorithms.

3. We obtain the input and output degree of each vertex.

4. We order the vertices from highest to lowest output degree, when two or more with the same output degree, sort them by the input degree, from low to high

5. We take v_1, and create the first set of community vertices $C_1 = C_1 \cup \{v_1\}$

   (a) For every vertex v_k,
   (b) If $v_k \in C_1 \vee C_2 \vee .... \vee C_{k-1}$ take the next vertex.
   (c) Otherwise we define a new set of Community vertices $C_k = C_k \bigcup \{v_j; v_j \in R(v_k), v_j no \in C_1 \vee C_2 \vee .... \vee C_{k-1}\}$†

6. We order the vertices from highest to lowest input degree, if there are two or more vertices with the same input degree, sort them by the output degree, from lowest to highest.

7. We take v_1, and created the first set of community vertices $C'_1 = C'_1 \cup \{v_1\}$

(a) For every vertex v_k,

(b) If $v_k \in C'_1 \vee C'_2 \vee .... \vee C'k - 1$ take the next vertex.

(c) Otherwise we define a new set of Community vertices $C'_k = C'_k \bigcup \{v_j; v_j \in R(v_k), v_j no \in C'_1 \vee C'2 \vee .... \vee C'_{k-1}\}$ ‡

8. We built $G_{OUT} = \bigcup_{i \in I} G_{OUT(C_i)}$ as the graph generated by each community sets, C_i obttained in †.

9. We built $G_{IN} = \bigcup_{j \in J} G_{IN(C_j)}$ as the graph generated by each community sets, C_j obtained in ‡.

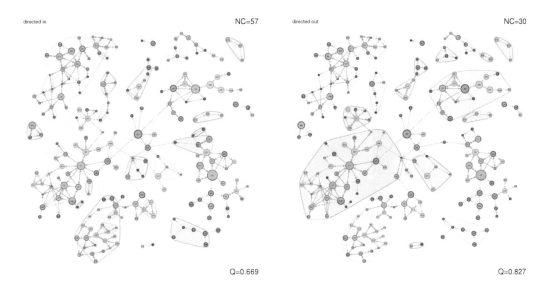

Figure 7. Graph In and Graph Out.

With Graph OUT we obtain the communities formed by highly emitters vertices and its receiving environment. An alteration in the demand for these highly emitter vertices, affects all vertices of the community.

With Graph IN we identify vertices receptors and the community that provide traffic vertices to that vertex. These vertices receivers will be sensitive to any variation in their contributors.

Here are some examples of the behavior of communities obtained by Graph IN and Graph OUT on the graph:

In the case of the subgraph of Figure 4, the communities obtained for these vertices coincide both using Graph IN as Graph OUT. This is a stable community. Thus, the changes of the offer at any of these vertices will mean little variations in the remaining community vertices. This case makes sense in our example, since there are degrees of the area of Arts, which are characterized as being highly vocational.

Regarding engineering, referred in Figure 5, we represent in Figure 8 the communities obtained using the algorithm Graph OUT. It allows us to detect vertices or groups of vertices

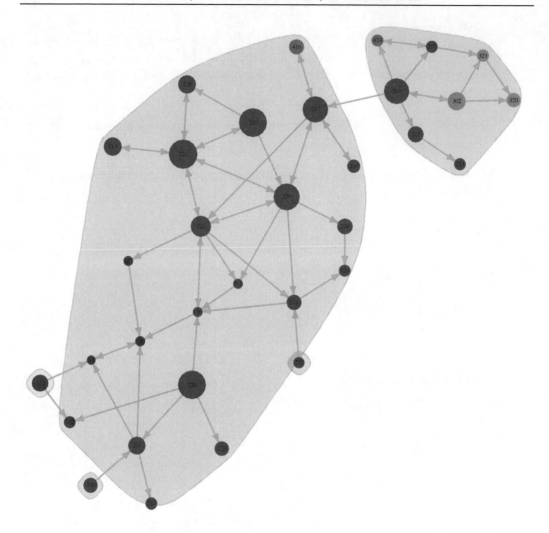

Figure 8. Communities subgraph for engineering. Graph OUT.

that only have outbound traffic. The modifications in the offer in these vertices would allow us to control the flows.

We can see that the vertex 203, corresponding to the degree of Architecture, is the main source of traffic in the upper right community. However, in the larger community, this role is distributed between the vertices 226, 229, 217 and 223, corresponding to Computers, Mechanical Engineering, Industrial Design and Industrial Engineering. These vertices have a much greater supply than its demand, which allows the sending of students to their second and third choices.

It remains to study the communities formed by isolated vertices, see the vertices 101, 240 or 223. In these vertices, a modification in its offer only affects a few vertices from another community, and their effect should be studied into greater detail in the future.

It should also be studied the role of vertices belonging to different communities, but those in where existing an edge in the graph. These would be vertices that serve as a bridgehead between two communities, but do not have sufficient entity to join communities.

The Graph IN algorithm provides us less information, but complements the one given by Graph OUT.

Using the Graph IN algorithm, we can detect vertices or groups of vertices that only have incoming traffic. The supply modification in these vertices would allow us to control some flows.

The edges that connect communities obtained with Graph IN, indicate that traffic is sent from a community to another whose offer is lower. In this way, we can detect secondary degrees related to the demand for large communities. See Figure 9 for the subgraph of Engineering.

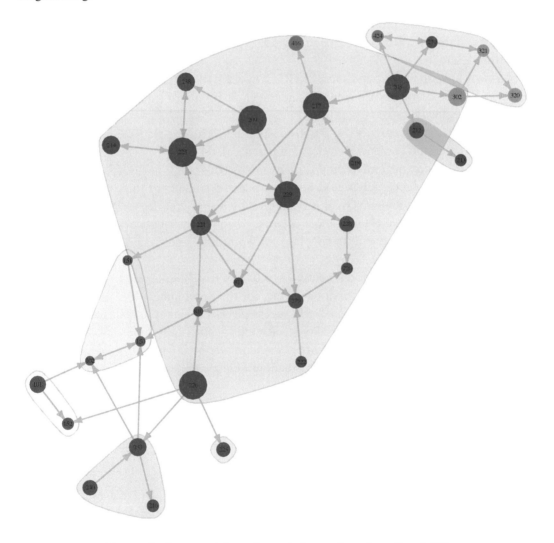

Figure 9. Communities subgraph for engineering. Graph IN.

Conclusion

We have proposed an algorithm to represent the students traffic between degrees in the process of access to S.U.P.E. The result is a directed graph of respectable size depending on each autonomous community and variable from year to year.

Once obtained this graph, we have proposed an algorithm that allows us to group the vertices of the graph in directed communities, with high modularity, that guarantees a good level of relationship between the vertices.

The algorithm has been applied to several regions and the results are similar and consistent with the problem being analyzed. Examples of the application of the algorithm and a brief interpretation of the results obtained are presented.

The algorithms are programmed using R Project, cite R, and can be found in the directions "http://www.upv.es/orgpeg/pub/CONSTRUCTION-OF-THE-GRAPH.r" the Algorithm 1, and in
"http://www.upv.es/orgpeg/pub/CONSTRUCTION-OF-DIRECTIONAL-COMMUNITIES.r" for the Algorithm 2, and the graph of Figure 1, Figure 2, Figure 6 and Figure 7, respectively:
"http://www.upv.es/orgpeg/pub/The-graph-for-a-system-250-deg-x-25000-stu.pdf"
"http://www.upv.es/orgpeg/pub/Metodos-comparados-1011.pdf"
"http://www.upv.es/orgpeg/pub/Metodos-comparados-0102.pdf"
"http://www.upv.es/orgpeg/pub/Metodos-comparados-03050608.pdf"
"http://www.upv.es/orgpeg/pub/Graph-In-and-Graph-Out-for-250-deg-x-25000-stu.pdf"

With this algorithm we obtain a more accurate vision of the students selection process, and this allows us to implement up new lines of work once studied and verified the algorithm.

In this sense, this algorithm can be used as a tool of *Academic Analytics* to assist those in responsible of the public university systems, for the development and modification of plans for distribution of the number of new access places to their systems, creation of new qualifications, etc. It would be necessary to develop specific tools aimed for this purpose and adapted to the final user. This requires the incorporation of specialized personnel and is beyond the scope of this work. Notwithstanding we hope to develop it in the future.

References

[1] Ahuja, R. K., Magnanti, T. L., and Orlin, J. B. Network flows. Upper Saddle River, Prentice Hall. 1993.

[2] Barabsi, A. L., Jeong, H., Nda, Z., Ravasz, E., Schubert, A., and Vicsek, T. (2002). Evolution of the social network of scientific collaborations. *Physica A: Statistical mechanics and its applications*, 311(3), 590-614.

[3] Boccaletti, S., Latora, V., Moreno, Y., Chavez, M., and Hwang, D. U.. Complex networks: Structure and dynamics. *Physics reports*, 424(4), 175-308.2006

[4] Bollen, Kenneth A. Structural equations with latent variables. John Wiley & Sons, 2014.

[5] Clauset,A. Newman, M. E. and Moore, C. Finding community structure in very large networks *Physical review E*, vol. 70, no. 6, p. 066111, 2004.

[6] Csermely, P. (2008). Creative elements: network-based predictions of active centres in proteins and cellular and social networks. *Trends in biochemical sciences*, 33(12), 569-576.

[7] Fortunato,S. Community detection in graphs, *Physics Reports*, vol. 486, no. 3,pp. 75-174, 2010.

[8] M. Girvan and M. E. Newman Community structure in social and biological networks *Proceedings of the National Academy of Sciences*, vol. 99, no. 12, pp. 7821-7826, 2002.

[9] Gross, J. L., and Yellen, J. (Eds.). Handbook of graph theory. Boca Ratn, CRC press.2004.

[10] Guàrdia Olmos, J., Peró M., Hervás A., Capilla R., Soriano-Jiménez, P.P. and Porras M. Factors related with the university degree selection in Spanish public university system. An structural equation model analysis.*Quality & Quantity*. Vol 48. N. 2, 541-557.2015

[11] Harary, F. (Ed.). (2015). A seminar on graph theory. Courier Dover Publications.

[12] Hervás, A. Guàrdia Olmos, J., Peró M., Capilla R., Soriano-Jiménez, P.P. A Structural Equation Model for Analysis of Factors Associated with the Choice of Engineering Degrees in a Technical University. *Abstract and Applied Analysis*. Vol 2014, 2014.

[13] Muñoz-Repiso, M. El sistema de acceso a la universidad en Espaa: tres estudios para aclarar el debate. Madrid. MEC. 1997.[The university access system in Spain: three studies to clarify the debate. Madrid. MEC. 1997.]

[14] Newman, M.E. Modularity and community structure in networks. *Proceeding of theNational Academy of Sciences*. 103 (23), 8577-8582. 2006

[15] Newman M. E. and Girvan M. Finding and evaluating community structure in networks *Physical review E*, vol. 69, no. 2, p. 026-113, 2004.

[16] Newman M. E. Finding community structure in networks using the eigenvectors of matrices *Physical review E*, vol. 74, no. 3, p. 036-104, 2006.

[17] Newman M. E. Fast algorithm for detecting community structure in networks,*Physical review E*, vol. 69, no. 6, p. 066133, 2004.

[18] Peró, M., Soriano-Jiménez, P.P., Capilla R., Guàrdia Olmos J. and Hervás A. Questionnaire for the assessment of factors related to university degree choice in Spanish public system: A psychometric study. *Computers in Human Behavior* Vol 47: pp. 128-138, 2015.

[19] Pons, P. and Latapy,M. Computing communities in large networks using random walks.*J. Graph Algorithms Appl.*, vol. 10, no. 2, pp. 191-218, 2006.

[20] The R Project for Statistical Computing. www.r-project.org. (5-2016)

[21] Raghavan,U. N. Albert,R. and Kumara. S. Near linear time algorithm to detect community structures in large-scale networks.*Physical Review E*, vol. 76, no. 3,p. 036-106, 2007.

[22] Simko, G. I., and Csermely, P. (2013). Nodes having a major influence to break cooperation define a novel centrality measure: game centrality. *PloS one*, 8(6), e67159.

Chapter 3

Validation of INCODE Framework for Assessment of Innovation Competency of Higher Education Students: A Multidimensional Technique for Affinity Diagram to Detect the Most Relevant Behaviours and Skills

Mónica Martínez-Gómez[1],*, Manuel Marí-Benlloch[1] and Juan A. Marin-Garcia[2]

[1]Departamento de Estadística e Investigación Operativa aplicadas y Calidad
Universitat Politècnica de Valencia, València, Spain
[2]ROGLE-DOE-Universitat Politécnica de Valencia, Valencia, Spain

Abstract

Innovation is a complex process that comprises several competencies. Innovation includes perception of opportunities, ideas generation and evaluation, action plans, cooperation and risk. It is considered one of the most competitive advantage in determining the success or failure of a company in the global market. The main purpose of this study is to apply an affinity diagram to detect the most relevant behaviours and skills that assessment innovation competency of higher education students. We used the INCODE-ICB-v5 questionnaire as the measurement instrumen for individual innovation. According this questionnaire, the capacities and skills that comprise innovation competency can be broken down into three categories: individual, teamwork and networking especified by European Qualifications Framework for Learning. The results show that affinity diagram have clustered capacities and skills of innovation competence into the three same categories (individual, interpersonal and networking) according to previous cuantitatives studies of the measurement model of the INCOME questionnaire. It provides an alternative cualitative mechanism for validating questionaires.

---

* Corresponding Author address: Email: momargo@eio.upv.es.

Keywords: innovation competency, multivariate analysis, correspondence analysis, INCODE questionnaire

INTRODUCTION

During the last few years the competences-centred perspective appears to be a very attractive option in the field of education, with the aim of developing students' skills by incorporating better academic education processes and allowing them a greatest success in the labour market. Within this perspective, the European Higher Education Area (EHEA) has been designed to include also transversal skills.

Transversal skills are the disciplines and capabilities that can be used for all professions. They are generic skills related to the integrated implementation of acquired attitudes, personality features, knowledge and values.

Currently, one of the prerequisites required by companies is that professionals improve their qualifications in transversal skills. These aspects are very much taken into account when hiring their partners and they are also very important when it comes to start up a company. At this point, innovation appears as a key competence in the business world. Being able to develop innovation skills is a must in our society. Innovation is important, both at the personal and at the organisational level. Innovation represents the strategic process for competitiveness and people are at the centre of this innovation process. Thus, people's training to develop the competence of innovation is a must for all companies which want to be competitive.

However, it is not easy to define innovation, since it implies the acquisition of different capabilities and capacities among which the following are noteworthy: creativity (generation of ideas, critical thinking, synthesis/reorganization ability), creative problem-solving (using new ideas to solve problems as a leader or entrepreneur), problem identification (clarify the real nature and the cause of the problems, search continuous improvement, collect information), independent thinking, be open to new ideas, focus on research, team work, forward-looking approach, among others and which have been discussed in different papers (Kairisto-Mertanen, L. and Mertanen, O., 2012; Marin-Garcia, J.A., Gonzalez-Ladrón de Cevara, F., 2011; Marin-Garcia, J.A., Pérez-Peñalver, M.J., Vidal-Carreras, P.I. and Maheut, J., 2012; Penttilä, T. and Kairisto-Mertanene, L., 2012).

There is a lot of bibliography written about innovation skills. The publications derived from international projects are among the most relevant studies about this matter: the Innovation Competencies Development (INCODE), where the IC Barometer has been developed.

Therefore, there is not one exclusive classification to group the different capabilities or characteristics that make up innovation (Berdrow, I. and Evers, F.T., 2010; Cerinšek, G. and Dolinsek, S., 2009; De Jong, J.P.J. and Kemp, R., 2003; Kessler, E.H., 2004; Marin-Garcia et al., 2011) and there is much debate on the instruments used in order to identify and validate the acquisition of innovation skills, which means a lack of knowledge on the efficiency of teaching and learning methods.

Figure 1. Model of innovation competence construction based on Penttilä et al. (2011; 2012).

A model specifically centred on innovation competencies will be followed in this project (Lehto, A., Kairisto-Mertanen, L. and Penttilä, T., 2011; Penttilä et al, 2011; Penttilä and Kairisto-Mertanene, 2012; Watts et al., 2012) and which has been reproduced in one of the more widely used instruments during the last few years to measure that competence: the Innovation Competencies Barometer (INCODE-ICv5 Barometer) (Marin-Garcia et al., 2011; Penttilä and Kairisto-Mertanene, 2012; Watts et al., 2012) which measures innovation through three dimensions of capacity and talent recommended by the European Qualifications Framework for Learning, following the model proposed by Penttillä et al., (2011; 2012): individual, interpersonal and network, so that the students' innovation levels can be assessed and methods to foster its development can be provided (Figure 1).

This questionnaire has been designed with a formative approach. However, some authors have proposed a validation as reflective measurement model (Watts el al. 2013; Räsänen, in review), that is, an approach in which the observed data are caused by unobservable variables or constructs using quantitative techniques, despite several experts believing that it was a formative approach (Marin-García et al., 2013) both for the first order model, for the individual, interpersonal and network dimensions as well as, subsequently, in the second order model to measure the innovation competence through the other three dimensions (Jarvis, C.B., MacKenzie, S.B. and Podsakoff, P.M., 2003; Marin-García, J.A. et al., 2011).

Purpose and Contributions of Present Study

The main purpose of this study is to apply an affinity diagram to detect the most relevant behaviours and skills that assessment innovation competency of higher education students.

Affinity diagraming is a powerful method for encouraging and capturing lateral thinking in a group environment (Burtner, May, Scarberry, LaMothe and Endert, 2013). Affinity-diagram activities also helps teams to group and link their collective thoughts into a clear and understandable structure (Kawakita, J. 1991). Affinity driagram is usually conducted using pens, sticky notes and whiteboard. However, in recent years, many studies have been conducted to development and build solutions to improve the effectiveness group brainstorming or affinity diagrams using electronic system (Alloway, 1997; Awasthi and Chauhan, 2012; Onwuegbuzie, Bustamante and Nelson, 2009; Santos, G. 2006; Widjaja, Yoshii and Takahashi, 2014).

Although qualitative research has been critiqued as too often lacking in scholarly rigor, nowadays many researchs contradict it (Gioia, Corley and Hamilton, 2012). We have been unable to locate any studies that have validated INCOME questionnarie with a multidimensional qualitative technique. So, there is a lack of empirical studies that have addressed affinity diagram to assessment the innovation competency of higher education students.

This study is structured as follows. First, we present the research methodology. Second, the results obtained. Finally, this chapter concludes with the main reflection of findings achieved in our analysis, their limitations and recommendations for further research.

METHODS

Participants

We can consider four possible methods of input data collection which are necessary to assess students. Rankings, ratings, BARS and BOS and parited comparisons (Dowdy et al., 2013, Hatzinger and Dittrich, 2012, Marin-Garcia et al., 2012). In this chapter we will focus on the second method.

The total sample was constituted by 918 students of a Massive Online Open Courses (MOOC) from a Spanish public university, who will complete one version of the questionnaire (INCODE-ICB-v5) classifying the 25 items related to innovation competencies. The items on this version of the questionnaire INCODE-ICB-v5 are in a different order with respect to the original questionnaire, where they are ordered in blocks: individual, interpersonal and network. Thus, a random organization of the items in the questionnaire prevents the bias of a certain cluster.

The respondents were then divided in two groups. The first group, made up of 458 students, was required to freely classify the 25 items of the questionnaire into four categories, which had to be labelled by them. The second group was constituted by 460 respondents, which were required to classify the same 25 items, but in this case they had to do it freely in a maximum of 10 categories, which they also had to label.

Thus, in principle, there are not any categories in which to classify the items and the respondents are completely free to express their criteria of association and similarity among the items on the questionnaire related to their perception about the concept of innovation.

Instrument

We selected the INCODE-ICB-v5 questionnaire (Watts et al., 2012; Marin-Garcia et al., 2011) which measures the innovation construct with a series of 25 questions, grouped into three categories (Individual, Interpersonal – teamwork – and Networking). Responses were given a score of between 1 and 5 (1 = major improvement needed; 5 = excellent).

Due to limitations of space, we are unable to list the items of the INCOD-ICB-v5 questionnaire in this chapter, but requests can be made via email for a copy of the questionnaire in either Spanish or English.

Methodology

Affinity diagramming is a technique used to externalize, make sense of, and organize large amounts of unstructured and dissimilar qualitative data.

In this study we develop qualitative explorative analysis as an optional tool to quantitative techniques for validating questionnaires via a distance-based affinity analysis, where individuals can group the proposed items of the questionnaire into the categories they think transversal innovation competence is better described. Our affinity diagramming process consists of two stages. First, the individuals group the items in the category they think is best. Then, we represent items in a perceptual map using correspondence analysis.

We used correspondence analysis as a multidimensional analysis to obtain the perceptual map.

The procedure allows the respondents to gain a perceived image related to a set of objects and interpret the dimensions of this space in attributes or dimensions interesting for the researcher, based on comparisons among objects. The aim is to understand the respondents' perceptions on the attributes of the study objects and chart the results on a perceptual map, transforming the similarity assessment among objects perceived by the respondents in distances between objects, being the researcher's art and science to interpret the dimensions of said perceptual map and assign it the relevant attributes (Hair et al., 1999).

In our case, the objects match up with items in the questionnaire and the purpose is to identify the respondents' judgment on the concept of innovation by searching the perceived attributes in the dimensions of the perceptual map obtained by correspondence analysis. In order to do this, the similarity comparisons among the items on the questionnaire carried out by the respondents shall be transformed in distances among said items which will be represented in the perceptual map.

RESULTS

In order to analyse the internal structure of the data collected from the questionnaire, we have used the Nonmetric Multidimensional Scaling (PROXSCAL). The procedure for this technique is the optimal position determination among objects. From a desired initial dimensionality, configurations are obtained by calculating the distances among objects and comparing the relationships observed against the relationships estimated with an adjustment for measurement. The configurations are a distribution of the set of objects (dots) on the coordinate axes which make up the dimensions and which can be represented on a "perceptual map." (Hair et al., 1999). Once the configuration is found, the distances among objects (d_{ij}) in the configuration are compared with the distance measurements (e_{ij}) of similarity judgements. These two distances are then composed using an adjustment for measurement, called stress measurement. The directions in which a higher adjustment can be obtained in order to reach a configuration with a satisfactory stress measurement and with the lowest possible dimensionality are set out below.

Kruskal's stress is a measure proposed to determine the adjustment of a model:

Kruskal's Stress = sqr $[(d_{ij} - e_{ij})^2 / (d_{ij} - d_m)^2]$

where,

d_m = the average distance on the map ($\sum d_{ij}/n$)

d_{ij} = distance obtained from the similarity data

e_{ij} = original distances provided by the respondents.

Practically all the individuals responded to all 25 items on the questionnaire, so any missing values are not due to the characteristics of an item, nor do they present a problem for the data collected as a whole.

The multiple correspondence analyses were carried out with two databases, one for four categories model and second for up to ten categories model.

In order to determine the similarity among items perceived by the respondents, the frequencies with which each item was classified into a category together with other item were calculated, that is to say, the number of times the respondents placed those two items into the same category. The absolute frequencies of similarity among items for each one of the groups are shown in Figures 2 and 3.

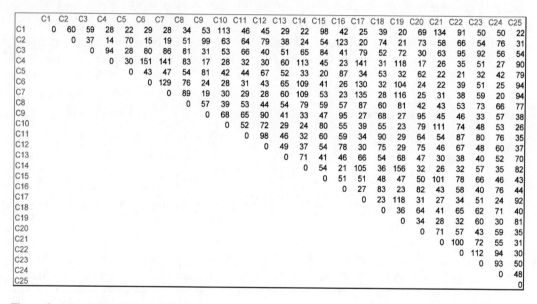

Figure 2. Association frequencies between items for the group of four categories.

	C1	C2	C3	C4	C5	C6	C7	C8	C9	C10	C11	C12	C13	C14	C15	C16	C17	C18	C19	C20	C21	C22	C23	C24	C25
C1	0	36	33	12	12	19	7	15	27	94	21	21	15	15	55	19	8	30	12	43	127	57	28	36	13
C2		0	17	16	53	18	13	33	67	24	32	43	23	13	26	78	19	40	10	68	29	24	17	52	20
C3			0	71	17	46	64	29	17	31	36	16	32	51	38	12	45	25	54	15	33	71	89	29	25
C4				0	24	92	87	53	14	11	23	19	34	75	30	11	100	16	85	14	11	34	55	14	45
C5					0	20	25	26	44	21	22	40	23	19	22	60	25	24	26	41	10	15	20	37	70
C6						0	78	60	24	18	20	29	35	61	35	16	81	32	69	21	15	29	38	22	41
C7							0	66	17	10	16	18	38	79	47	9	109	13	87	17	11	22	50	11	42
C8								0	70	46	65	61	64	93	64	59	103	63	88	64	47	53	65	66	79
C9									0	66	63	93	53	45	55	109	48	72	39	109	54	49	43	73	55
C10										0	44	48	28	14	49	19	9	36	11	47	94	51	26	37	10
C11											0	93	50	21	31	40	14	51	15	28	16	44	42	41	31
C12												0	78	47	60	80	45	78	46	79	51	65	50	73	55
C13													0	52	36	17	43	51	46	26	14	27	31	28	37
C14														0	36	12	77	29	130	26	21	23	42	21	47
C15															0	23	47	39	33	27	58	42	51	26	26
C16																0	16	51	13	61	17	26	17	56	34
C17																	0	16	89	20	12	25	45	15	50
C18																		0	22	37	31	44	27	48	34
C19																			0	23	14	23	51	17	50
C20																				0	43	23	23	42	28
C21																					0	60	37	39	15
C22																						0	77	76	31
C23																							0	50	32
C24																								0	47
C25																									0

Figure 3. Association frequencies between items for the group of ten categories.

Multidimensional Scaling: Four Categories Model

In table 1, we can see that stress and measurements for adjustment indicate the efficiency with which the distances of the solution get closer to the original distances.

In the four categories model, each of the stress statistics measure the mismatch of the data, so stress values are close to 0 (Normalized raw stress 0.01947). On the other side, the explained dispersion and Tucker's consistency coefficient measure the adjustment of the model and, in our case, these measurements for adjustment get closer to value 1 (Tucker consistency coefficient 0.99022). All of this means we are before an excellent solution.

Table 1. Measures of stress and adjustment for the model 4 categories

Normalized raw stress	.01947
Stress-I	.13955[a]
Stress-II	.31831[a]
S-Stress	.03720[b]
Told dispersion (D.A.F.)	.98053
Tucker consistency coefficient	.99022

Note: PROXSCAL minimizes normalized stress raw. a. Optimal scaling factor = 1.020
b. Optimal scaling factor = .996.

The perceptual map obtained for the first group of respondents of four categories is shown below in Figure 4.

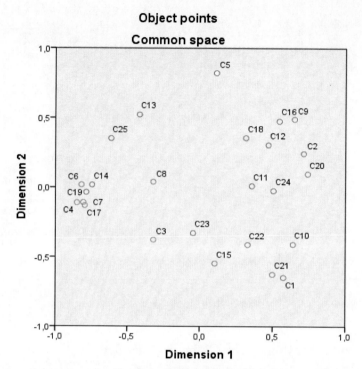

Figure 4. Perceptual Map for the model of four categories.

The chart on Figure 4 shows the two first dimensions related to the 25 items of the questionnaire. The items have been coloured according to the three components supposed for innovation (individual, interpersonal and network). As we can observe in the map, the items corresponding to the individual component, in blue, are clearly grouped, except item 23. The items of the interpersonal component, in green, present a very compact grouping made up of items (4, 6, 7, 17) with a clear approach to items 14 and 19 which belong to the networking component. On the other hand, item 8 is isolated from the rest of items, in the centre of the common space and items 3 and 15 seem to be approaching to other items of the individual component (22 and 23). Meanwhile, the items of the networking component 13, 25 and 5 are to be found in the left top space creating a dispersed association indeed far from the rest of items.

Multidimensional Scaling: Model With Up to Ten Categories

In Table 2, we can see stress values and measurements for adjustment, which indicate the efficiency with which the distances of the solution get closer to the original distances. In the up to ten categories model, each of the stress statistics measure the mismatch of the data, so stress values are close to 0 (Normalized raw stress 0.03222). The explained dispersion and Tucker's consistency coefficient measure the adjustment of the model and get closer to value 1 (Tucker consistency coefficient 0.98376). All of this means we are before an excellent solution.

The map on Figure 5 shows the two first dimensions related to the 25 items of the questionnaire. As in the case of four categories, items have been coloured according to the three components supposed for innovation. As we can observe in the map, the items corresponding to the individual component, in blue, are clearly grouped, except item 23. The items of the interpersonal component, in green, present two groupings, except item 8 which is isolated. On one hand, the grouping (7, 4, 6, 17) shows a clear approach to items 14 and 19 which belong to the networking component. On the other hand, a second grouping (3, 15, 22) which is associated to item 23 of the individual component. Finally, the items of the networking component 13, 25 and 5 are to be found in the top space creating a dispersed association indeed far from the rest of items in other components (individual and interpersonal).

Table 2. Measures of stress and adjustment for the model up to 10 categories

Normalized raw stress	.03222
Stress-I	.17949[a]
Stress-II	.42954[a]
S-Stress	.06183[b]
Told dispersion (D.A.F.)	.96778
Tucker consistency coefficient	.98376

Note: PROXSCAL minimizes normalized stress raw. a.Optimal scaling factor = 1,033.
b. Optimal scaling factor = ,992.

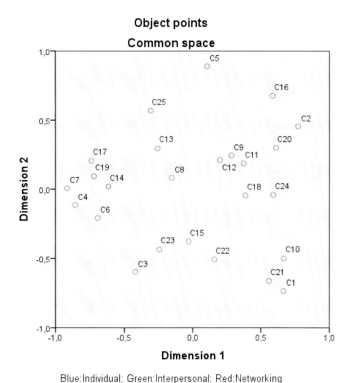

Figure 5. Perceptual Map for for the model up to 10 categories.

Conclusion

The purpose of our study is to evaluate if affinity diagram can be used to validate questionnaries as an alternative to the cuantitative techniques. Results shows the cualitative validation of INCODE questionnaire to assessment the innovation competency of university students.

The internal consistency of theorics components of innovation is high in both models, for four and up to ten categories. Besides, the instrument has been validated as reflective and formative measurement model with cuantitatives techniques, empirical results of multidimensional scale confirms the structure of these three componets (individual, interpersonal and networking), although some unsettled items were detected, in particular items C3, C8, C15 y C23 because they are away from their theorical related items.

But in any case, the internal mesuarement of the innovation competecency is maintained. These findings are useful for researchers since they add the first sample in which the validation of a competency is developt with qualitative techniques and results are according with other qualitatives techniques, like Strustural Model Equation (SEM) or Partial Least Squared (PLS).

Results over the technical caracteristics of the instrument, suggest real application for the improvement of measurement innove competency.

There were of course, limitations to this study. As stated previously, we used a student sample with a specific questionnaire and the generalization to other questionnaire, or population, should be proved with specific data.

Acknowledgments

This chapter has been written with financial support from tree Projects:

FINCODA: "Project 554493-EPP-1-2014-1-FI-EPPKA2-KA" (The European Commission support for the production of this publication does not constitute an endorsement of the contents which reflects the views only of the authors, and the Commission cannot be held responsible for any use which may be made of the information contained therein).

Project PIME 2015-2016 A/09 "Evaluación de los indicadores del comportamiento innovador en el alumno universitario", at the Universitat Politècnica de València (Spain).

Project GVA/2016/004 de la Conselleria d'Educació, Investigació, Cultura i Esport de la Generalitat Valenciana (Spain).

References**

Alloway, A. (1997). Be prepared with an affinity diagram. *Quality Progress, 30*(7), 75–77. Retrieved from http://www.scopus.com/inward/record.url?eid=2-s2.0-0031190504&partnerID=40&md5=6c32d2393401409441e40ac990f6ef4a.

Awasthi, A. and Chauhan, S.S. (2012). A hybrid approach integrating affinity diagram, AHP and fuzzy TOPSIS for sustainable city logistics planning. *Applied Mathematical Modelling*, 36 (2012) 573–584.

Berdrow, I., Evers, F. T. (2010) «Bases of competence: an instrument for self and institutional assessment». Assessment and Evaluation in Higher Education, Vol. 35, nº. 4, pp. 419-434.

Burtner, R., May, R., Scarberry, R., LaMothe, R. and Endert, A. (2013). Affinity+: Semi-Structured Brainstorming on Large Displays. *POWERWALL: International Workshop on Interactive, Ultra-High-Resolution Displays, Part of the SIGCHI Conference on Human Factors in Computing Systems, CHI '13 Extended Abstracts on Human Factors in Computing Systems (CHI EA '13)*, 1–6.

Cerinšek, G. and Dolinsek, S. (2009). Identifying employees' innovation competency in organisations». International *Journal of Innovation and Learning*, 6, (2), 164-177.

De Jong, J. P. J., Kemp, R. (2003) «Determinants of CoWorkers' Innovative Behaviour: An Investigation into Knowledge Intensive Services». International Journal of Innovation Management, Vol. 07, nº. 02, pp. 189-212.

Gioia, D. A., Corley, K. G. and Hamilton, A. (2012). Seeking qualitative rigor in inductive research: Notes on the Gioia methodology. *Organizational Research Methods*. doi: 10.1177/1094428112452151.

Hair, J. F., Hult, G. T., Ringle, C. M. and Sarstedt, M. (2013). A Primer on Partial Least Squares Structural Equation Modeling (PLS-SEM). Thousand Oaks: Sage.

Jarvis, C.B., MacKenzie, S.B. and Podsakoff, P.M. (2003). A Critical Review of Construct Indicators and Measurement Model Misspecification in Marketing and Consumer Research. *Journal of Consumer Research*, 30, 199-218.

Kairisto-Mertanen, L. and Mertanen, O. (2012). Innovation pedagogy- a new culture for education. Revista de Docencia Universitaria Volume: 10, Issue 1; ISSN 1887-4592, p.p.: 67-86. http://red-u.net/redu/index.php/REDU/article/view/333.

Kawakita, J. (1991). *The original kj method*. Tokyo: Kawakita Research Institute.

Kessler, E. H. (2004). Organizational innovation: A multi-level decision-theoretic Perspective. *International Journal of Innovation Management, 8, (3), 275-295*.

Lehto, A., Kairisto-Mertanene, L. and Penttilä, T. (2011). Towards innovation pedagogy. A new approach to teaching and learning for universities of applied sciences. Turku University of Apllied Sciences.

Lizasoain, L. and Joaristi, L. (2012). Las nuevas tecnologías y la investigación educativa. El análisis de datos de variables categoriales. *Revista Española de Pedagogía 251, 111- 130*.

Marin-Garcia, J. A., Aznar-Mas, L. E. and González-Ladrón deGevara, F. (2011). Innovation types and talent managment for innovation. *Working Papers on Operations Management,2, (2), 25-31*.

Marin-Garcia J.A.; Pérez-Peñalver, María José.; Vidal-Carreras, PI.; Maheut, J. (2012). *How to assess the innovation competency of higher education students*. Proceedings of the 7th International Conference on Industrial Engineering and Industrial Management, p.p.: 920-928.

Marin-Garcia, J. A., Perez-Peñalver, M. J. and Watts, F. (2013). How to assess innovation competence in services: The case of university students. Direccion y Organizacion (50), 48-62. Retieved from: http://www.revistadyo.com/index.php/dyo/article/viewFile/431/451.

Onwuegbuzie, A. J., Bustamante, R. M. and Nelson, J. A. (2009). Mixed Research as a Tool for Developing Quantitative Instruments. *Journal of Mixed Methods Research, 4, (1), 56-78*.

Penttilä, T. and Kairisto-Mertanene, L. (2012). Innovation competence barometer ICB - a tool for assessing students' innovation competences as learning outcomes in higher education, in INTED2012 Conference. 5th-7th March 2012, pp. 6347-6351.

Räsänen, M. (in review). Validation of innovation competence barometer.

Santos, G. 2006. Card sort technique as a qualitative substitute for quantitative exploratory factor analysis. *Corporate Communications 11(3), 288-302.*

Watts, F., Garcia-Carbonell, A. and Andreu Andrés, M. A. (2013). Innovation competencies development: Incode barometer and use guide. Turku: Turku University od Applied Sciences.

Watts, F., Marin-Garcia, J.A., Garcia-Carbonell, A. and Aznar-Mas, L.E. (2012) Validation of a rubric to assess innovation competence. Working Papers on Operations Management 3: 61-70.

Widjaja, W., Yoshii, K., Haga, K. and Takahashi, M. (2013). Discusys: Multiple user real-time digital sticky-note affinity-diagram brainstorming system. *Procedia Computer Science*, 22(0), 113–122. http://doi.org/10.1016/j.procs.2013.09.087.

In: Modeling Human Behavior: Individuals and Organizations   ISBN: 978-1-53610-197-3
Editors: L. Jódar Sánchez, E. de la Poza Plaza et al.   © 2017 Nova Science Publishers, Inc.

Chapter 4

EVALUATION OF M-LEARNING AMONG STUDENTS ACCORDING TO THEIR BEHAVIOUR WITH APPS

Laura Briz-Ponce[1,*], *Anabela Pereira*[2],
Juan Antonio Juanes-Méndez[1]
and Francisco José García-Peñalvo[1]

[1]University of Salamanca, Salamanca, Spain
[2]University of Aveiro, Aveiro, Portugal

ABSTRACT

The present chapter has the goal to provide some insights regarding the current use of mobile technologies for learning. This research was conducted at University of Salamanca and University of Aveiro and took into account the collaboration of 518 students from both universities.

The main results indicate that the students are very willing to use m-learning and there is a relationship between the use of mobile devices (frequency of use of Tablet) and the use of Apps with the global evaluation of m-learning by students. However, most part of students still reported an unawareness and a lack of necessity of these instruments, which brings into light that it is necessary to support and promote the use of these technologies with a curricular and educational purpose by institutions and universities.

Keywords: higher education, m-learning, mobile devices, m-health, students

INTRODUCTION

Mobile technologies using for learning have become an upward trend in our society. The rapid spread of accessing mobile devices among students has caused they have been used for many purposes. Overall, thanks to the emergence of Apps, which are software programs that could run on mobile devices as Smartphones or Tablets to provide them with additional

---

[*] Corresponding Author address; Email: laura.briz@usal.es.

functionalities. One of the potential uses of these new technologies is using them as educational tools. There are some researches about this issue, but there is still a gap regarding the real impact and benefits that could improve in the students' learning. Also, there are some challenges and barriers that it is necessary to overcome, as for example technical problems (Alrasheedi et al., 2015; Green et al., 2015; Toktarova et al., 2015; Handal et al., 2013; Székely et al., 2013), the support of the Institution of University (Alrasheedi et al., 2015; Alden, 2013; Ashour et al., 2012; Park et al., 2012; Lea and Callaghan, 2011), the lack of skills to use them (Haffey et al., 2014; Ferreira et al., 2013; Ozdalga et al., 2012; Fadeyi et al., 2010), the need of a pedagogical goal of the Apps (Ferreira et al., 2013; Handal et al., 2013; Székely et al., 2013; Ashour et al., 2012; Davies et al., 2012) or even the need of regulation of Apps that may cause a lack of trust on the efectiveness of them as instructional instruments for learning (Martínez-Pérez et al., 2015; Haffey et al., 2014; Khatoon et al., 2013; Visvanathan et al., 2012).

On the other hand, the different benefits are been also reported by different authors (Toktarova et al., 2015; Archibald et al., 2014; Ling et al., 2014; Ventola, 2014; Al-fahad, 2009; Hussain and Adeeb, 2009) standing out among these advantages the ubicuity or possibility to use the mobile devices anywhere, the flexibility and the possibility to access information easily.

Therefore, the potentional uses of mobile devices and Apps are still under study. This chapter tries to cover this gap in order to analyse more deeply the current different students' uses for learning and the role that these tools could have over them.

METHODS

Method

The method used for this research was a non-experimental descriptive-correlational transaccional investigation, using a mixed methodology (quantitative and qualitative) with a deductive reasoning. We will collect the information from different variables and then, they will be correlated taking into account the independent variables (predictors) and the dependent variables (criteria).

Variables

The variables used for this research are detailed in this section. The table 1 describes them differentiating between dependent and independent variables. The results section will provide information regarding the relation between both types of variables. In our case, we only have one dependent variable, called VGLOB and measures the level of acceptance of m-learning between students.

The predictor variables considered for this study will be the frequency of use of participants with Smartphone and Tablet, the type of device that participants use the most to download Apps, the Characteristics that participants consider more important to download Apps and finally the type of Apps that participants use more frequently.

Table 1. Summary of Variables used in the reseach

Type	ID	Description	Values
Independent Variable	FREQ_{SMP}	Indicates how many daily hours use the participants the Smartphone	<1 h/day
			From 1 to 2 h/day
			From 3 to 4 h/day
			>4 h/day
			No use
	FREQ_{TAB}	Indicates how many daily hours use the participants the Tablet	<1 h/day
			From 1 to 2 h/day
			From 3 to 4 h/day
			>4 h/day
			No use
	DEV	Indicates what is the device most used to download Apps	Smartphone
			Tablet
			Smartphone and Tablet
			None
	N°Apps_{SMP}	Describes the number of Apps downloaded with Smartphone	From 1 to 10
			From 11 to 20
			From 21 to 30
			>30
			None
			N/A
	N°Apps_{TAB}	Describes the number of Apps downloaded with Tablet	From 1 to 10
			From 11 to 20
			From 21 to 30
			>30
			None
			N/A
	CHAR_{APPS}	Reports the characteristics more important to download Apps. It could be	Security/Privacy
			Content
			Usability
			Accesibility
			Data Connexion
			Recommendation
			Developer Information
			None
	TYPE_{Apps}	Reports the type of Apps that the participants used more frequently. It could have the values	Entertainment
			News
			Social Networks
			Mail
			Games
			Medical Apps
			Educational Medical Apps
			Other
			None
Dependent Variable	VGLOB	Indicates the total evaluation of using m-learning among participants	Numerical

Participants

The number of participants of this study was 518. As it is shown on Table 2, 96,9% of participants owned a mobile device (Smartphone or Tablet). Besides, most part of participants were women, were studying medicine and were within the range from 18 to 25 years old. The most popular operating system was Android for both Smartphone and Tablet.

Table 2. Descriptive Statistics of Students' Profile

Variable	Basic Profile Characteristics		
	Description	Frequency	%
Grade	Medicine	222	26,9
	Nursing	105	18,2
	Biomedical Sciences	136	29,8
	Physioterapy	37	8,1
	Doctorate	5	1,1
	Psychology	13	2,8
Sex	Male	113	21,8
	Female	405	78,2
Age	From 18 to 25 years	487	94,0
	From 26 to 35 years	19	3,7
	From 36 to 45 years	9	1,7
	+ 55 years	3	0,6
Mobile Device	Only Smartphone	206	39,8
	Only Tablet	24	4,6
	Smartphone and Tablet	272	52,5
	None	16	3,1
Operating System Smartphone	iOS (iPhone)	93	18,0
	Android	365	70,5
	Windows8	15	2,9
	N/A	38	7,2
	Do not know	7	1,4
Operating System Tablet	iOS (iPad)	83	16,0
	Android	164	31,7
	Windows 8	37	7,1
	Otros	7	1,4
	N/A	223	43,1
	Do no t know	4	0,7

Instruments

The instrument used for this resarch was a survey of 53 questions distributed in two parts. The first one was formed by 19 items to collect information from participants' profile. The second one was formed by 34 items designed according to the model proposed by Venkatesh et al. (2003) to unify the different theories of behaviour use and the acceptance of technology.

In our survey we added as well two more constructs related with the reliability and the Recommendation of new technologies for m-learning.

The survey was distributed from May to June 2014 at University of Spain and October and December 2015 at University of Aveiro and University of Coimbra in Portugal.

All the data was computerized using SPSS program (V.21) in order to obtain the descriptive statistics and the main results of the study.

RESULTS

This research presents the results of the students' use of mobile devices and Apps and how their profile could influence in the final evaluation of m-learning.

Use of Mobile Devices

The data collected from participants gave us information regarding how students were using mobile devices and the frequency of daily use. We differentiated between the use with Smartphones and the use with Tablets. According to the results, there is around 48,3% of participants that use the Smartphone from 1 to 2 hours per day and the tablet is used by 32,6% of students.

The Figure 1 represents the box plot chart considering the frequency of use with Smartphone and the median of global evaluation of m-learning. As it is shown in it, it seems that the median of evaluation of m-learning is very similar among participants.

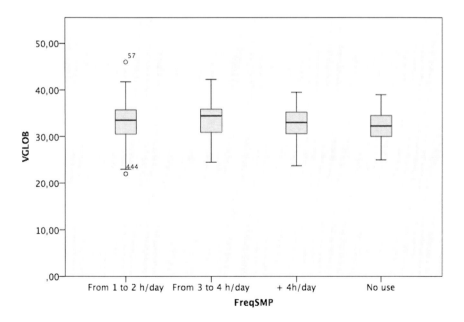

Figure 1. Global evaluation of M-learning taking into account the frequency use of Smartphone.

We want to estimate the degree and correlation of relationship of these variables (Freq$_{SMP}$ and the VGLOB). As we are comparing one nominal variable with a numerical variable, it is necessary to check if they fulfil the needed requirements to use parametric techniques (Field,

2000). We use Kolmogorov-Smirnov Test to check the normality condition and we obtain in all cases that $\rho>0,05$ so we can assume that the variables are normal. Besides, we perform as well the test of Levene to assess the homogeneity and we obtain as well that $\rho>0,05$. Therefore, we can use the parametric variance technique to contrast the variables. In this case, the null hypothesis is that there is no relationship between the frequency use of Smartphone and the global evaluation. The results (F=0,582 and $\rho=0,676$) reveal that at $\alpha=0,05$, there is no evidence enough to fail to reject the null hypothesis that there is no relationship between both variables.

Then, we perform the same analysis with frequency of use of Tablet. The results are also showed in Figure 2.

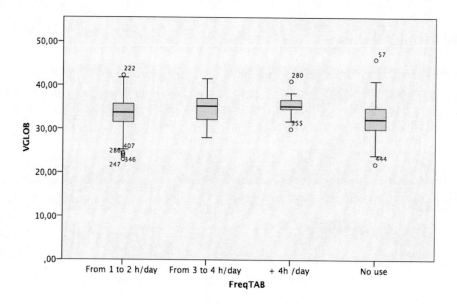

Figure 2. Global evaluation of M-learning taking into account the frequency use of Tablet.

We carried out again the same process, obtaining that they fulfiled the requirements to use parametric technique (the variables are normal and they are homogeneous). The null hypothesis wass that there is no relationship between the frequency of use with Tablets and the global evaluation of m-learning. In this case, according to the results (F=9,722 and $\rho=0,000$), we could suggest that at 0,05 level of significance there is evidence enough to reject the null hypothesis and consider there is a relationship between both variables.

Use of Apps

According to the results, students were mainly using the Smartphones to download Apps (77,8%) and 47,1% of them were using the Tablet. Besides, 55,6% of participants downloaded from 1 to 10 Apps last month with Smartphone and 37,8% with Tablets.

Then, we checked the normality requirement for all variables and we obtained that all of them could be considered as normal ($\rho >0,05$) and all fulfil the homogeneity test ($\rho >0,05$) so it is possible to use parametric techniques in all cases. The null hypothesis in all cases is that there is no relationship between the predictor variable and the global evaluation of m-learning. The table 3 shows the output data obtained with the suitable technique applied. In all cases, we obtain that at 0,05 level of significance, there is enough evidence to reject the null hypothesis that consider both variables independents and we could suggest that among students, there is a relationship between the number of Apps downloaded with the Smartphone, with the Tablet, the type of device used and the global evaluation of m-learning.

Table 3. Results of contrasting test used between the use of Apps and global evaluation of m-learning

Predictor Variable	Dependent Variable	Technique	Result	
			F	ρ
N°Apps$_{SMP}$	VGLOB	Analysis of Variance	4,285	0,000
N°Apps$_{TAB}$	VGLOB	Analysis of Variance	6,398	0,000
DEV	VGLOB	Analysis of Variance	4,199	0,006

In addition, we also obtained information of the relevant characteristics that students took into account when they downloaded an App. In fact, according to the results, the ranking of the factors are shown in Figure 3.

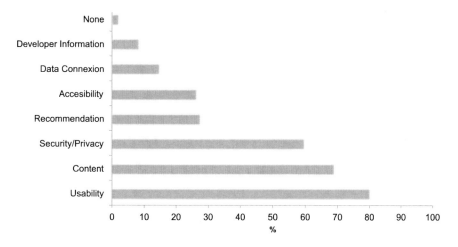

Figure 3. Ranking of relevant factors to download Apps.

We performed the same analysis as well, checking the normality and homogeneity test. In this case, the variable VGLOB did not fulfil the requirement of normality (ρ>0,05) with the independent variable CHAR$_{APPS}$ for Security/Privacy, Content and Usability.Therefore, it was necessary to use the non-parametric tecnique U-Mann Whitney. On the other hand, for the rest of values, the normality wass positive and the test of homogeneity showed that the variable CHAR$_{APPS}$ for accesibility (F=0,948, ρ=0,331), data connexion (F=0,938, ρ=0,333), Recommendation (F=2,498, ρ=0,115), developer information (F=0,022, ρ=0,883) and none of those characteristics (F=0,251, ρ=0,617) are all homogeneous so in all these cases, it was possible to use a parametric test (t Student). The Table 4 represents the outcome data obtained with the different techniques applied. The null hypothesis was that there is no relationship between the independent variable and the global evaluation of m-learning (VGLOB). The results suggested that there is no evidence enough to reject the null hypothesis ($p>0,05$) for the participants that selected Accessibility, Recommendation and Developer information as relevant factors to download apps. On the contrary, according to the results and at 0,05 level of significance, there is evidence enough to reject the null hypothesis considering that participants who have selected Security/Privacy, Content, Usability, Data Connexion and none of them as relevant factors could give more scores to the evaluation of m-learning.

Table 4. Results of contrasting test used between the relevant factors to download Apps and global evaluation of m-learning

Independent Variable	Technique	Results	
		t/Z	p
$CHAR_{APPS\ security/privacy}$	U-Mann Whitney	-3,195	0,001
$CHAR_{APPS\ Content}$	U-Mann Whitney	-2,279	0,023
$CHAR_{APPS\ Usability}$	U-Mann Whitney	-2,443	0,015
$CHAR_{APPS\ Accessibility}$	t Student	-0,352	0,725
$CHAR_{APPS\ Data\ Connexion}$	t Student	-3,999	0,000
$CHAR_{APPS\ Recommendation}$	t Student	0,305	0,760
$CHAR_{APPS\ developer\ Information}$	t Student	-0,203	0,839
$CHAR_{APPS\ None}$	t Student	3,338	0,011

Finally, we analysed the type of Apps that the participants used the most. Figure 4 shows that Apps of Social Networks and Entertainment are the ones most used. In this case, the educational Apps were only used by 20,1% of participants. This type of apps was considered as the most interesting to contrast with the global evaluation of m-learning. Therefore, we applied again the parametric technique t Student (we checked previously normality and homogeneity test) and according to the results ($t=-3,696$, $p=0,000$), we can suggest that there is enough evidence to reject the null hypothesis and accept the alternative one that indicates that there is a relationship between the participants who have used educational Apps and the global evaluation of m-learning.

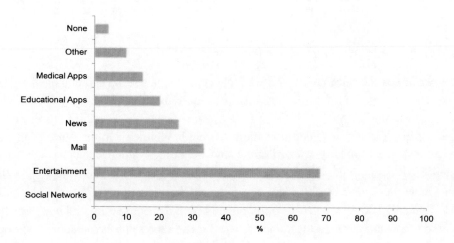

Figure 4. Type of Apps that participants used the most.

Challenges

Other important information obtained within this research was the students' reasons of no using educational Apps. This data could be very valuable in order to analyse the main barriers and challenges that the institutions or organizations should get over in order to adopt m-learning as a new curricular technique.

The results indicate that no necessity and unawareness as the main factors for not using them, so it is important to establish a pedagogical goal of this type of Apps in order that participants will find them useful and promote their use and their access in order to make them more popular.

Table 5. Students' reasons for no using educational Apps

Reason	Frec	%	Reason	Frec	%
No necessity	73	17,9%	No access	5	1,2%
Unawareness	38	9,2%	Utility	4	0,9%
Not enough quality	20	4,8%	No technical skills	4	0,9%
Better Books or computer	22	5,3%	No interest	3	0,7%
N/A	11	2,6%	Storage of device	2	0,5%
No trust	8	1,9%	Few apps	1	0,2%
Price	6	1,4%	No time	1	0,2%

DISCUSSION AND CONCLUSION

The results of this research provide some insights about the use of Apps in Higher Education and the most important factors that could drive to give more evaluation of using m-learning. We used a cohort of Spanish and Portuguese students and the results indicate that 96,9% of participants owned a mobile device (Smartphone or Tablet), which is also confirmed by other researchs to highlight the rapid expand of these devices among students (Chen et al., 2015; Briz-Ponce et al., 2014a, 2014c).

Besides, we obtain that there is a relationship between the frequency of use of Tablet and the global evaluation of M-learning by students. Also, there is a relationship between participants that have downloaded more apps during the last month and the assesment of m-learning. Regarding the use of Apps, we obtained that participants who have selected Security/Privacy, Content, Usability, Data Connexion and none of them as relevant factors could give more scores to the evaluation of m-learning. Finally, participants who have used educational Apps scored m-learning higher than the ones who have not used them.

These results may contribute to define new behaviour patters to use mobile technologies as the one performed with women in Education (Briz-Ponce, Juanes-Méndez and García-Peñalvo, 2016) and allow focus on the main challenges to adopt these new type of technologies: No necessity and unawareness. Other researches analyse also the advantages or disadvantages of using these new technologies (Briz-Ponce et al., 2014c; Chu et al., 2012) or even the potential instructional uses of these tools for learning (Briz-Ponce, Juanes-Méndez, García-Peñalvo, et al., 2016; Briz-Ponce and García-Peñalvo, 2015; Briz-Ponce and Juanes-

Méndez, 2015) bringing to light that it is necessary to deal with different barriers and claiming that the leadership of Universities and Organizations must support them and provide an awareness-raising campaign about the use of educational Apps. This challenge will allow a special continuos education and promote life long learning, which is one of the purposes of the organizations. There are some guides that could be useful for them in order to adopt these changes and modify the behaviour in their Institutions (Michie et al., 2014).

Finally, the promotion and incentivation of individuals, self regulation and their soft skills may contribute to enhance the usage of mobile devices and Apps and capacitate individuals to be prepared for the new digital world.

REFERENCES

Al-fahad, F. N. (2009). Students' Attitudes and Perceptions Towards the Effectiveness of Mobile Learning in King Saud University, Saudi Arabia. *The Turkish Online Journal of Educational Technology*, *8*(2), 111–119.

Alden, J. (2013). Accomodating mobile learning in college programs. *Journal of Asynchronous Learning Networks*, *17*(1), 109–122.

Alrasheedi, M., Capretz, L. F. and Raza, A. (2015). A Systematic Review of the Critical Factors for Success of Mobile Learning in Higher Education (University Students' Perspective). *Journal of Educational Computing*, *52*(2), 252–276. http://doi.org/10.1177/0735633115571928.

Archibald, D., Macdonald, C. J., Plante, J., Hogue, R. J. and Fiallos, J. (2014). Residents' and preceptors' perceptions of the use of the iPad for clinical teaching in a family medicine residency program. *BMC Medical Education*, *14*, 174. http://doi.org/10.1186/1472-6920-14-174.

Ashour, R., Alzghool, H., Iyadat, Y. and Abu-Alruz, J. (2012). Mobile phone applications in the university classroom: Perceptions of undergraduate students in Jordan. *E-Learning and Digital Media*, *9*(4), 419–425. http://doi.org/10.2304/elea.2012.9.4.419.

Briz-Ponce, L. and García-Peñalvo, F. J. (2015). An Empirical Assessment of a Technology Acceptance Model for Apps in Medical Education. *Journal of Medical Systems*, *39*(11), 176. http://doi.org/10.1007/s10916-015-0352-x.

Briz-Ponce, L. and Juanes-Méndez, J. A. (2015). Mobile Devices and Apps, Characteristics and Current Potential on Learning. *Journal of Information Technology Research*, *8*(4), 26–37. http://doi.org/10.4018/JITR. 2015100102.

Briz-Ponce, L., Juanes-Méndez, J. A. and García-Peñalvo, F. J. (2016). The role of Gender in Technology Acceptance for Medical Education. In M. M. Cruz-Cunha, I. M. Miranda, R. Martinho and R. Rijo (Eds.), *Encyclopedia of E-Health and Telemedicine* (p. Vol II, pp. 1018–1032). Hershey, PA: IGI Global.

Briz-Ponce, L., Juanes-Méndez, J. A. and García-Peñalvo, F. J. (2014a). A systematic review of using mobile devices in medical education. In B. D. ierra-Rodriguez J.-L.,Dodero-Beardo J.-M. (Ed.), *Proceedings of 2014 International Symposium on Computers in Education (SIIE* (pp. 205–210). Logroño: Institute of Electrical and Electronics Engineers Inc. http://doi.org/10.1109/SIIE.2014.7017731.

Briz-Ponce, L., Juanes-Méndez, J. A. and García-Peñalvo, F. J. (2014b). First Approach of mobile applications study for medical education purposes. In *Proceedings of the Second International Conference on Technological Ecosystems for Enhancing Multiculturalit* (pp. 647–651). New York, NY, USA: ACM New York.

Briz-Ponce, L., Juanes-Méndez, J. A. and García-Peñalvo, F. J. (2014c). Analysis of Mobile devices as a support tool for professional medical education in the University School. In *6th International Conference on Education and New Learning Technologies - EDULEARN14* (pp. 4653–4658). Barcelona: IATED Academy.

Briz-Ponce, L., Juanes-Méndez, J. A., García-Peñalvo, F. J. and Pereira, A. (2016). Effects of Mobile Learning in Medical Education: a Counterfactual Evaluation. *Journal of Medical Systems*, *40*(6), 1–6.

Chen, B., Seilhamer, R., Bennet, L. and Bauer, S. (2015). Students' Mobile Learning Practices in Higher Education: A Multi-Year Study. *EDUCAUSE Review*.

Chu, L. F., Erlendson, M. J., Sun, J. S., Alva, H. L. and Clemenson, A. M. (2012). Mobile computing in medical education: opportunities and challenges. *Current Opinion in Anaesthesiology*, *25*(6), 699–718. http://doi.org/10.1097/ACO.0b013e32835a25f1.

Davies, B. S., Rafique, J., Vincent, T. R., Fairclough, J., Packer, M. H., Vincent, R. and Haq, I. (2012). Mobile Medical Education (MoMEd) – how mobile information resources contribute to learning for undergraduate clinical students - a mixed methods study. *BMC Medical Education*, *12*(1), 1. http://doi.org/10.1186/1472-6920-12-1.

Fadeyi, A., Desalu, O. O., Ameen, A. and Adeboye, A. N. M. (2010). The reported preparedness and disposition by students in a Nigerian university towards the use of information technology for medical education. *Annals of African Medicine*, *9*(3), 129–34. http://doi.org/10.4103/1596-3519.68358.

Ferreira, J. B., Klein, A., Freitas, A. and Schlemmer, E. (2013). Mobile learning: Definition, uses and challenges. In L. A. Wankel and P. Blessinger (Eds.), *Cutting-edge Technologies in Higher Education* (pp. 47–82). Emerald Group Publishing Limited. http://doi.org/ 10.1108/S2044-9968(2013)000006D005.

Field, A. (2000). *Discovering statistics using SPSS for Windows*. Londres: SAGE Publications.

Green, B. L., Kennedy, I., Hassanzadeh, H., Sharma, S., Frith, G. and Darling, J. C. (2015). A semi-quantitative and thematic analysis of medical student attitudes towards M-Learning. *Journal of Evaluation in Clinical Practice*, *21*(5), 925–930. http://doi.org/10.1111/jep.12400.

Haffey, F., Brady, R. R. W. and Maxwell, S. (2014). Smartphone apps to support hospital prescribing and pharmacology education: a review of current provision. *British Journal of Clinical Pharmacology*, *77*(1), 31–8. http://doi.org/10.1111/bcp.12112.

Handal, B., Macnish, J. and Petocz, P. (2013). Academics adopting mobile devices : The zone of free movement. In *30th ascilite Conference 2013 Proceedings* (pp. 350–361).

Hussain, I. and Adeeb, M. A. (2009). Role of mobile technology in promoting campus-wide learning environment. *Turkish Online Journal of Educational Technology*, *8*(3), 48–57.

Khatoon, B., Hill, K. B. and Walmsley, a D. (2013). Can we learn, teach and practise dentistry anywhere, anytime? *British Dental Journal*, *215*(7), 345–347. http://doi.org/10.1038/ sj.bdj.2013.957.

Lea, S. and Callaghan, L. (2011). Enhancing health and social care placement learning through mobile technology. *Journal of Educational Technology and Society*, *14*(1), 135–145.

Ling, C., Harnish, D. and Shehab, R. (2014). Educational Apps: Using Mobile Applications to Enhance Student Learning of Statistical Concepts. *Human Factors and Ergonomics in Manufacturing*, *24*(5), 532–543. http://doi.org/10.1002/hfm.

Martínez-Pérez, B., de la Torre-Díez, I. and López-Coronado, M. (2015). Experiences and Results of Applying Tools for Assessing the Quality of a mHealth App Named Heartkeeper. *Journal of Medical Systems*, *39*(11), 1–6. http://doi.org/10.1007/s10916-015-0303-6.

Michie, S., Atkins, L. and West, R. (2014). *The Behaviour Change Wheel Book - A Guide To Designing Interventions*. UK: Silverback Publishing.

Ozdalga, E., Ozdalga, A. and Ahuja, N. (2012). The smartphone in medicine: a review of current and potential use among physicians and students. *Journal of Medical Internet Research*, *14*(5), e128. http://doi.org/10.2196/ jmir.1994.

Park, S. Y., Nam, M. and Cha, S. (2012). University students' behavioral intention to use mobile learning: Evaluating the technology acceptance model. *British Journal of Educational Technology*, *43*(4), 592–605. http://doi.org/10.1111/j.1467-8535.2011.01229.x.

Székely, A., Talanow, R. and Bágyi, P. (2013). Smartphones, tablets and mobile applications for radiology. *European Journal of Radiology*, *82*(5), 829–836. http://doi.org/10.1016/j.ejrad.2012.11.034.

Toktarova, V. I., Blagova, A. D., Filatova, A. V. and Kuzmin, N. V. (2015). Design and Implementation of Mobile Learning Tools and Resources in the Modern Educational Environment of University. *Review of European Studies*, *7*(8), 318–324. http://doi.org/10.5539/res.v7n8p318.

Venkatesh, V., Morris, M. G., Davis, G. B. and Davis, F. D. (2003). User Acceptance of Information Technology: Toward a Unified View. *MIS Quarterly*, *27*(3), 425–478.

Ventola, C. L. (2014). Mobile devices and apps for health care professionals: uses and benefits. *P and T: A Peer-Reviewed Journal for Formulary Management*, *39*(5), 356–64.

Visvanathan, A., Hamilton, A. and Brady, R. R. W. (2012). Smartphone apps in microbiology-is better regulation required? *Clinical Microbiology and Infection*, *18*(7), E218–E220. http://doi.org/10.1111/j.1469-0691.2012. 03892.x.

Chapter 5

ASSESSING UNIVERSITY STAKEHOLDERS ATTRIBUTES: A PARTICIPATIVE LEADERSHIP APPROACH

Martín A. Pantoja[1], María del P. Rodríguez[1] and Andrés Carrión[2]

[1]Universidad Nacional de Colombia, Facultad de Ingeniería y Arquitectura, Departamento de Ingeniería Industrial, Campus La Nubia, Manizales, Colombia
[2]Universitat Politècnica de Valencia, Centro de Gestión de la Calidad y del Cambio, Valencia, Spain

ABSTRACT

In this chapter, the relationship between leaders and stakeholders is analysed. Specifically, the point of interest is the role played by the stakeholders in modifying leaders behaviour. Stakeholders influence is expressed by their attributes (power, legitimacy and urgency), and the expressions of participative leadership are consult, autocracy, joint decision and delegation. After a review of the questions a model is proposed, and with the aim of applying it in a specific context, a questionnaire is presented, validated and applied. With a relational approach and from a subjective perspective, perceptions of a sample of leaders from public universities in Manizales (Colombia) were collected. A first group of constructs was formed, including the university stakeholders attributes mentioned above. A second group of constructs collects their relevance. Reliability of constructs was measured using Cronbach alpha, and its values indicate that is feasible to measure effectively the proposed constructs. It is concluded that the questionnaire has the internal consistency and reliability for assessing the university stakeholders' attributes. In the analysis, it has been considered that stakeholders are determined by the organizational context and that relevance of the attributes are the result of leaders´ perceptions.

Keywords: university stakeholders, participative leadership, attributes assessment

LEADERSHIP AND STAKEHOLDERS

Referring to leadership necessarily implies talking about relevance and influence of stakeholder (groups of interest) in the organization. Forgetting this is equivalent to deny the organizations' systemic nature and even that of leadership itself, as interdependence among internal and external stakeholders is a fact, as mentioned by Vroom and Jago (1995) or Porter and McLaughlin (2006), for whom leadership is contextual. Organizations are immersed in a frame formed by internal and external actors, who have the capacity of influencing organization processes. Leadership in general and leaders with managerial responsibilities in particular, are not free from this influence. The authors of this paper have studied the relationship between stakeholders and leaders with managerial responsibilities in a Higher Education environment (Pantoja et al., 2015).

From a wide perspective, there exists an interchange relationship between the operation of internal processes and the influence of the different actors (Mintzberg, 1983) or stakeholders (Freeman, 2001). Their basic attributes (power, legitimacy and urgency) (Mitchell et al., 1997) are the influential instruments. Definitions of these attributes were taken from Mitchell et al. (1997):

- Power is the relationship among social actors in which one stakeholder can get another social actor to do something that would not have otherwise done
- Legitimacy is a generalized perception or assumption that the actions of an entity are desirable, proper, or appropriate within some socially constructed system of norms, values, beliefs and definitions.
- Urgency is the degree to which stakeholder claims call for immediate attention.

One of these internal processes is exercising participative leadership in which a leader, in a managerial position, gives some degree of autonomy and participation to its co-operators to allow them to influence in the decision making process (Yukl, 2010). Nevertheless, relation between stakeholders' attributes relevance and the way leadership is manifested (specially participation) hay received little attention by the academic literature (Mitchell et al., 1997; Schneider, 2002; Myllykangas et al., 2010).

Leadership is not alien to the organization's nature, and this is in a comprehensive context (Katz and Kahn, 1977; Osborn et al., 2002). According to Porter and McLaughlin (2006) context, which has not been studied according to its relevance for the understanding of leadership phenomenon, is formed by organization's groups of interest (Freeman, 2001) who interact forming interchange social networks (Homans, 1961; Blau, 1964).

Depending on the value given by the leader to the attribute of a specific stakeholder, he assigns a stakeholder relevance and, depending on this may or not become object of interest and attention. These attributes are perceived by the leader, processed and finally, according to Katz and Kahn (1977), affect the leader behaviour. In this way stakeholder become actors not only with power, but also with the legitimacy and urgency required to influence the leader's behaviour, as the communicate their expectations to the leader affecting his "focus person" behaviour (Katz and Kahn, 1977). In consequence, leadership in organizations is based in the stakeholders and oriented to the stakeholders.

To study this situation, it is convenient to develop a questionnaire and validate the constructs that permit to evaluate stakeholders' attributes from the leaders' perceptions view point, and in an individual analysis.

The context of leadership action, and specially the actors (stakeholders), has received little attention and those researches that consider this as a relevant question uses a theoretical approach not implemented in practice. Congruent and systematic models are required to develop applied research, allowing identifying and describing in a better way the phenomenon of the interchange relationships between leaders and the related stakeholders.

The model used in this analysis is represented in Figure 1. The unit of study is the person, in our case a leader in a public university, with managerial responsibilities. His perceptions about how stakeholders are influencing his actions are the subject of interest of the analysis, joint with the relevance of the different attributes (power, legitimacy and urgency). This influence modifies the participative leadership expressions (consult, autocracy, joint decision and delegation).

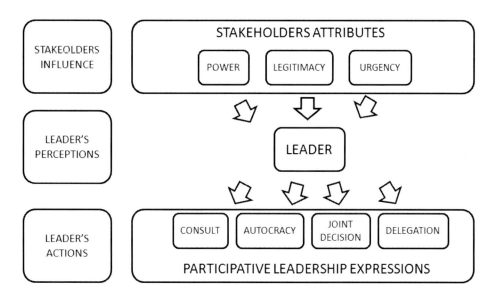

Figure 1. Stakeholders' attributes influence over the leader.

Ten stakeholders were considered in this analysis, as relevant influences present in the University. This ten groups were divided in two categories, internal and external stakeholders.

Initially, internal stakeholders were: University top management, professors, research groups and students. External stakeholders were: companies, community, financing bodies, government, alumni and environmental organizations. In a first round of contacts with participants an additional internal stakeholder was identified as potentially relevant and included in the study: the administrative staff. Finally eleven stakeholders were included in the analysis.

The stakeholder government refers to the Colombian State Government, and its role, relevant in any case, is especially important to consider in this study as we are working with public universities.

In this framework, a questionnaire to explain influential relationships between stakeholders' attributes and participative leadership expressions is a relevant tool.

QUESTIONNAIRE AND PILOT TEST

The objectives of this questionnaire are, first to give structure to a theoretical discussion modelling the above-mentioned relationships, second to allow proposing a model based on the identified relationships and third to illustrate the application of the proposed model for a specific context.

The questionnaire used to asses stakeholder attributes was designed to be used with an interviewer. Questions are closed, with a verbal scale of four levels, similar to others used in different researches on leadership. (Reche et al., 2008; Delgado et al., 2011). The scale advances from null (zero points) to high (three points), to reflect the influence of each attribute or the frequency in influence attempts and was selected to avoid the risk of having a tendency to the central value in scales with an odd number of levels.

The target population was formed by leaders with managerial responsibilities in Public Universities in Manizales (Colombia). Each interviewed is asked to rate the level on influence the different stakeholders have over their acting according to the attributes considered.

Part of the questionnaire is presented in figure 2.

#	Grupo de interés universitario	Atributo	Escala de Influencia				Frecuencia de los intentos de influencia			
			Alta 3	Media 2	Baja 1	Nula 0	Alta 3	Media 2	Baja 1	Nula 0
1	Directivas de la Universidad	Poder	☐	☐	☐	☐				
		Legitimidad	☐	☐	☐	☐	☐	☐	☐	☐
		Urgencia	☐	☐	☐	☐				
2	Profesores	Poder	☐	☐	☐	☐				
		Legitimidad	☐	☐	☐	☐	☐	☐	☐	☐
		Urgencia	☐	☐	☐	☐				
3	Grupos de Investigación	Poder	☐	☐	☐	☐				
		Legitimidad	☐	☐	☐	☐	☐	☐	☐	☐
		Urgencia	☐	☐	☐	☐				
4	Empresas	Poder	☐	☐	☐	☐				
		Legitimidad	☐	☐	☐	☐	☐	☐	☐	☐
		Urgencia	☐	☐	☐	☐				
5	Estudiantes	Poder	☐	☐	☐	☐				
		Legitimidad	☐	☐	☐	☐	☐	☐	☐	☐
		Urgencia	☐	☐	☐	☐				
6	Sociedad en general	Poder	☐	☐	☐	☐				
		Legitimidad	☐	☐	☐	☐	☐	☐	☐	☐
		Urgencia	☐	☐	☐	☐				

Figure 2. Questionnaire structure and items.

To check the validity of the questionnaire and to avoid overcharge the target population with an excessive number of surveys, a pilot test was run with 38 former managers (of different levels) from the three public universities of Manizales. The positions they have occupied are different (Rector, Vice-rector, Dean, Head of Department, Curricular Area Manager, Research Manager, ...) to obtain a good representativity, even considering that this was a pilot test.

RESULTS AND DISCUSSION

Two types on orthogonal construct were defined in the database. The first one defines each attribute as a construct; the second defines each stakeholder as a construct. A total of fourteen constructs were considered (three attributes plus eleven stakeholders).

To check the internal consistency and reliability of the instrument (the questionnaire) Cronbach Alphas were computed. Values over 0.5 are desirable and over 0.7 preferred (Helmstadter, 1964; Nunnally and Bernstein, 1999).

Table 1 shows the results obtained. Constructs corresponding to the stakeholders attributes have acceptable Cronbach Alpha values, with a especially good value for urgency. This result indicates the the instruments allows evaluating the three attributes in the different stakeholders considered with good reliability.

Table 2. Cronbach Alphas for Attributes constructs

Power	0.758
Legitimacy	0.693
Urgency	0.799

Table 3 presents the results corresponding to Cronbach Alphas for Stakeholders constructs. It includes two sets of values. The first one shows the results obtained when considering inside the construct the three attributes and the frequency of the influencing attempts. The second one shows the results considering in the construct only the three attributes.

The values obtained indicate that in general the level of coherence and reliability is acceptable or good, except for the stakeholders University top management and (to some point) government, which presents values under the desirable level.

Results confirm the presence of these eleven groups of interest in the University environment. These groups were suggested in different papers as Duque Oliva (2009), Vallaeys et al. (2009) and Rodríguez Fernández (2010). Those stakeholders with lower coherence can be understood as affected by contextual factors (Osborn et al., 2002; Porter y McLaughlin, 2006). Even considering the work by Vieira (2013), there was a change in the hierarchy of the groups of interest, with a relevant increase in the importance of the stakeholder administrative staff, absent in Vieira (2013) but included by Caballero Fernández et al. (2007).

Table 3. Cronbach Alphas for Stakeholders constructs

Stakeholders	Attributes + Frequency	Attributes
University top management	0.443	0.277
Professors	0.689	0.550
Research groups	0.749	0.728
Students	0.749	0.669
Companies	0.802	0.721
Community	0.736	0.652
Financing bodies	0.828	0.850
Government	0.479	0.512
Alumni	0.819	0.803
Environmental organizations	0.783	0.791
Administrative staff	0.903	0.930

The third column in Table 3, illustrates the effect of excluding the variable frequency of the influential attempts. Results are different in this column with reference to the second. In those cases where second column value is greater than the value in the third column, we can interpret that variable frequency enhances the clarity in the perception of the corresponding stakeholder by the leaders interviewed.

CONCLUSIONS

The particular context in each organization determines which stakeholders can be relevant and influential, according with the perception of the actors object of interest in each case. From the individual analysis level, and specifically from the perception of interviewed leaders, the proposed questionnaire is a valid instrument to indentify the stakeholders present in the public universities of Manizales (Colombia).

The statistical analysis indicates that the coherence and reliability of the constructs defined are adequate. This instrument can be used in further studies that go in deep in the analysis of the relationships among leaders and stakeholders, according to the model proposed in Figure 1.

REFERENCES

Antonakis, J., et al., Methods for studying leadership, In The nature of leadership by Antonakis, J., Cianciolo, A.T. y Sternberg, R. J., pp. 48-70 Sage Publications, Thousand Oaks (2004).

Blau, P. M., Exchange and power in social life, Wiley, New York (1964).

Caballero Fernández, G., J. M. García Vásquez, y M. A. Quintas Corredoira, La importancia de los stakeholders de la organización: un análisis empírico aplicado a la empleabilidad del alumnado de la universidad española, Investigaciones Europeas de Dirección y Economía de Empresas. [The relevance of stakeholders in the organization: an empirical analysis applied to employability of Spanish universities students. European Researches in Business Economy and Management].

Delgado, M. L. Las comunidades de liderazgo de centros educativos. Educar, ISSN: 2014-8801, (en línea), (48), 9-21, 2012. [Leadeship Communities in Education Centers].

Duque Oliva, E. J., La gestión de la universidad como elemento básico del sistema universitario: una reflexión desde la perspectiva de los stakeholders, Innovar, ISSN: 2248-6968, 19( ), 24-42, 2009. [University Management as a basic element of H.E. System: a review from stakeholders' perspective].

Freeman, R. E. A stakeholder theory of the modern corporation. Perspectives in Business Ethics Sie, 3, 144. 2001.

Helmstadter G. C. Principles of psychological measurement. New York: Appleton-Century-Crofts, 1964.

Homans, G. C., Social behavior: Its elementary forms, Harcourt, Brace and World, New York (1961).

Katz, D. y Kahn, R. L. Psicología social de las organizaciones, Trillas, México (1977). [Organizational Social Psychology].

Mintzberg, H., Power in and around organizations, Prentice Hall, Englewood Cliffs (1983).

Mitchell, R. K., B. R. Agle, y D. J. Wood, Toward a theory of stakeholder identification and salience: defining the principle of who and what really counts, *Academy of Management Review* 22(4), 853-886 (1997).

Myllykangas, P. J. Kujala y H. Lehtima Ki. Analyzing the essence of stakeholder relationships: What do we need in addition to power, legitimacy, and urgency? *Journal of Business Ethics96* (August), 65-72 (2010).

Nunnally, J. and Bernstein, I. (1999). Teoría psicométrica. México: Trillas. [Psychometric Theory].

Osborn, R. N., J. B. Hunt y L. R. Jauch, Toward a contextual theory of Leadership, *The Leadership Quarterly*, 13, 797-837 (2002).

Pantoja, Martín A., Rodríguez, María del P., Carrión, Andrés. Diseño de un Cuestionario para Valorar los Atributos de Grupos de Interés Universitarios desde un Enfoque de Liderazgo Participativo. *Formación Universitaria* Vol. 8(4), 33-44. 2015. [Design of a Questionnaire to Assess University Stakeholders Attributes from a Participative Leadership Approach].

Porter, L. W. y G. B. McLaughlin, Leadership and the organizational context: Like the weather?, *The Leadership Quarterly*: 17 (6), 559-576 (2006).

Reche, M. P. C., Delgado, M. L. and Martínez, T. S. Evaluación de la representación estudiantil en la Universidad desde un enfoque de género: diseño de un cuestionario. Enseñanza and Teaching: Revista interuniversitaria de didáctica, (26), 137-164, 2008. [Evaluation of Students representation in University from a gender approach: design of a questionnaire].

Rodríguez Fernández, J. M., Responsabilidad Social Universitaria: Del discurso simbólico a los desafíos reales. In Responsabilidad Social Universitaria by Cuesta González, M., Cruz Ayuso, C. y Rodríguez Fernández, J. M (Ed.), pp 3-24 Netbiblo. (2010). [University Social Responsibility: from the Symbolic Discourse to the real challenges. In University Social Responsibility].

Schneider, M., A Stakeholder Model of Organizational Leadership, *Organization Science*, 13(2), 209-220 (2002).

Vallaeys, F., de la Cruz, C., y Sasia, P. (2009), Responsabilidad Social Universitaria. Manual de Primeros Pasos, 1ª edición, Mc Graw Hill Interamericana Editores S.A. de C.V., México (2009). [University Social Responsibility. First Steps Handbook].

Vieira, J. A., The socially responsible management in Colombian public universities. Case study: the research function in public universities at Manizales (Colombia), Tesis de Doctorado, Université de Rouen Laboratoire du NIMEC. Rouen-France (2013).

Vroom, V. H. y A. G. Jago, Situation effects and levels of analysis in the study of leader participationThe Leadership Quarterly, 6 (2), 169-181 (1995).

Yukl, G., Leadership in organizations, seventh edition. Prentice Hall, New Jersey (2010).

Chapter 6

INTERVENTION PROGRAMME FOR PHARMACY OFFICE PREVENTING METABOLIC SYNDROME: IMPROVING THE POPULATION'S QUALITY OF LIFE BY MODELLING ITS BEHAVIOR

María del Mar Meliá Santarrufina[*]
and Fernando Figueroa[†]*, PhD*

Departamento de Tecnología de la Alimentación y Nutrición,
Universidad Católica de Murcia, Murcia, Spain

ABSTRACT

Changes in diets and lifestyles in past decades have been accelerated by various factors, like economic development, industrialisation, urbanisation and globalisation of markets. Economic development has played an important role in these achievements by facilitating education and health policies for most of the population. Despite economic prosperity, this has been accompanied by an increased incidence of chronic diseases.

The main objective of this chapter consists in generating strategies that help prevent, diagnose, control and treat metabolic syndrome through an intervention programme. Metabolic syndrome is defined as a set of metabolic disorders related to cardiovascular risk factors and predictors of diabetes development. This study proposes a non-pharmacological comprehensive intervention model used in pharmacy offices to study the prevalence of metabolic syndrome by considering a group of urban patients, adults aged over 20 years. It also aims to promote better a quality of life for patients given the consequent social impact due to increased productivity and reduced costs.

Keywords: prevention of chronic diseases, metabolic syndrome, pharmacy

---

[*] mariamelia@redfarma.es.
[†] ffigueroa@ucam.edu.

1. INTRODUCTION

Changes in diets and lifestyles over the past three decades have accelerated as a result of various factors including, among others, economic development, industrialisation, urbanisation and globalisation of markets. Populations with nutritional deficit have reduced, while another part of the population has grown in size and weight. Average life expectancy has increased on average by between 25-30 years, major infectious diseases have been eradicated, and infant mortality has drastically decreased.

Economic development has directly and indirectly played an important role in all these achievements through education and health policies that affect most of the population.

Despite economic prosperity, the incidence of chronic diseases has increased. Improved living standards with greater availability of food can promote the appearance of negative impacts, such as inappropriate eating habits, less physical activity, plus other harmful habits like drinking alcohol and smoking tobacco. This has had a major impact on populations' health and nutritional status, with a corresponding increase in chronic diseases related to diet, especially in certain population groups.

The set of metabolic disorders related to cardiovascular risk factors and predictors of diabetes development is commonly known today as metabolic syndrome (MS). The term MS was introduced by the World Health Organization, (WHO) in 1988 as a diagnostic entity with defined criteria [1]. According to the International Diabetes Federation (IDF), [2] people with MS are 3 times at more risk of suffering a stroke or heart attack compared to people without it, and are twice as likely to die of an event of this type.

For many years it has been known that diet is of crucial importance as a risk factor of MS, and recent scientific publications have shown that diet modifications may improve risk factors of MS [3]. Traditional diets based largely on plant food have been quickly replaced with a high-fat and high-calorie diet that mainly consists of foods of animal origin.

Dietary changes affect large human groups in different regions and countries, and should be modified in those people with MS or who are at risk of developing it by guiding them to a diet low in saturated fat, trans fats and cholesterol, and by reducing the intake of simple sugars and eating more fruit, vegetables and cereals.

The Mediterranean diet (MD) [4], which is simply the way how people who live on the Mediterranean coast eat, contains a large portion of these dietary recommendations: low-carbohydrate foods with low glycaemia index, and intake of fibre, soya, fruit and vegetables, and low saturated and trans fat, and cholesterol. Among the many beneficial health properties of the MD, the following are highlighted: type of fat that characterises olive oil, fish and nuts, proportions of key nutrients that certain recipes, cereals and vegetables contain as a basis for dishes and meat, which are rich in micronutrients, and is the result of eating fresh vegetables and condiments.

Low-carbohydrate diets can help control weight and blood pressure, and are able to improve insulin sensitivity and to reduce cardiovascular risk [5]. However it should be noted [6] that the carbohydrate type in our diet is important, such as rye, wheat, oats and potatoes. A diet rich in fibre from non-purified cereals makes insulin resistance difficult and thus favours lower MS prevalence.

The clear correlation between obesity and MS suggests that obese people resort more to medical care than normo-weight subjects, which results in increased public healthcare spending. It is estimated that obesity is responsible for between 1-3% of public health spending, except for the USA, where it lies between 5-10% [7]. It has been found in several countries that health spending on an obese person is more than 25% than on someone of normal weight [8]. Obesity adversely affects the production of most countries. Indeed in the USA it represents 1% of its GDP, and is 4% in China [9].

A delay between onset of obesity and health problems should be further noted as obesity has increased in previous decades, and current economic and immediate future costs will increase. For example by UK Foresight [10], which refers to the health costs related to obesity, in 2007 it was estimated that such costs would be 70% higher in 2015, and 2.4 times higher by 2025.

According to the NECP ATP III, the central goal of treatment with diet plans to treat MS [11] is that patients acquire a healthy lifestyle by eliminating factors of environmental and modifiable risk, which can be achieved by adapting diet, exercise and losing weight.

2. JUSTIFICATION OF WORK

Maintaining a healthy weight depends greatly on the control of food mismatches, so the key role lies in nutritional counselling and health responsibility is assumed by patients. Dieticians must motivate those patients who need to control their body mass index by reducing total calorie intake, increasing physical activity and making some nutritional changes by replacing calories from saturated fats, simple carbohydrates and monounsaturated fats with polyunsaturated fats of the omega-3 series.

One important facet of professional dieticians is to assume this multiple, clinical, psychological function, and accompanying patients with MS to make changes in their lifestyle that involve both alterations in eating styles and physical activity. The success of an MS intervention is conditioned by the dietician's ability to transmit these concepts and to make the intervention programme understood.

In pharmacy offices (PO), pharmacists act as a bridge between the patient and medication as they occupy the ideal position to provide optimum access to health care and to improve drug treatment. In fact many people with MS have no knowledge about it, so it is of the utmost importance that may are referred to a doctor for treatment to minimise complications.

Accordingly, PO concur in various circumstances to optimise the primary pharmaceutical care service. Firstly, the need to carry out nutritional interventions to correct food maladjustments, as stated above, may determine that an individual has or may have, cardiovascular disease and diabetes, among others. Secondly, professionals with proper training in nutritional interventions and the material means to conduct proper interventions are frequently offered non-scientific and media criteria.

While age, gender and genetic vulnerability are not modifiable elements, many of the risks associated with age and gender can be ameliorated. Such risks include the above-mentioned factors and are related to lifestyle, such as diet, physical inactivity, smoking and drinking alcohol, biological factors (dyslipidaemia, hypertension, obesity and hyperinsulinaemia) and, finally, social factors, which are a complex mix of cultural and socio-economic parameters, as well as other environmental elements that interact. Since MS is a largely preventable disease, it is considered that the primary public health prevention approach is the most economical and sustainable action to address the epidemic of disorders worldwide.

The main objective of this intervention is to generate strategies that help to prevent, diagnose, control and treat MS by developing a comprehensive non-pharmacological intervention model to study the prevalence of MS in urban patients, adults aged over 20 years. It aims to promote a better quality of life for patients given the consequent social impact due to increased productivity and reduced costs.

A key part of the intervention is to justify and convince patients why they should change their lifestyle. Patients must assume that the benefits of making the requested sacrifices can only be seen in diseases that may occur in the future.

3. Materials and Methods

Our intervention addresses the population recorded in the computer system of a PO in Valencia (Spain). For its natural surroundings and proximity, the population can be considered representative of two nearby neighbourhoods in the city of Valencia, with a total number of inhabitants of 23,337 people according to the 2014 census.

The primary source of information was the patient data available at the PO NIXFARMA 9.0.9 recorded in a computer system. A file with 3,016 patients for the 1997-2015 period was considered. Later years will be added in the future.

The sample to be analysed will be grouped according to age (over 20 years), gender, level of education, physical activity and previous pharmacotherapy. The International Diabetes Federation (IDF) will be adopted [12] as the MS diagnosis criteria. The methodology will be divided into four phases: preliminary, informative, assessment, training/counselling and clinical.

3.1. Preliminary Stage

The careful preparation of this preliminary phase is critical for the project's success and largely depends on meticulous preparation. A potentially large number of patients will be consulted, from whom we will obtain consent to be included in our analysis. Estimates have indicated some 1,900 individuals could respond, who should be adequately motivated to participate in the programme, which will logically entail time, a survey, and considerable expense.

A sample of 3,016 patients, enrolled in the programme management to be extended for the duration of the study, will be stratified by age, gender, level of education, previous drug treatment and physical activity using the NIXFARMA computer system, (9.0. 9). Therefore this stage will be divided into four activities: promotional programming, sample classification into groups, preparation and piloting the survey.

Promotional programming. It is very important to motivate the potential patients to enrol in the project. It is essential to publicise the nutritional counselling service offered. For this purpose, different promotional approaches will be used: posters and flyers announcing the service, and social networks Facebook and Twitter from the PO website.

Classification shown in categories. The 3,016 registered patients correspond, as stated above, to a record which began in 1997 with no other criteria other than pharmaceutical care. So it is necessary to update and rank patients into categories from the information recorded in the PO. The estimated number of potential respondents is 1,900, who we already know and can be classified according to them receiving medical, hypertension and cholesterol treatment, or not.

Next a database will be generated with the Nixfarma software tool and new patients will be added. For this purpose, Nixfarma has been contacted to support the use of the computer system in the specific and necessary functions in this project phase which are not common in the PO.

Preparation and pilot survey. It is necessary to incorporate preliminary survey piloting into this stage with a reduced sample of 10-15 patients before starting the survey.

3.2. Information Phase

Having completed the preliminary phase already and obtained the necessary material and information for patients, it will be disclosed to motivate and recruit the patients who will participate in the campaign. The duration and intensity of the campaign are both important; if it is short, it will not reach everyone; if it is too long, it may become trite. Therefore, flyers must be available on the PO counter, posters must be seen in the PO, and information must be posed in the social networks, etc, which all must come over strongly, but be reasonably short.

During this period we will telephone the patients with the information obtained from Nixfarma in the previous phase to offer them this service.

3.3. Assessment Phase, Training and Advice

This third phase is subdivided into two: evaluation, training and advice.

Evaluation. A phase when a newsletter will be delivered to the patients who positively responded to the information phase so they can provide their consent to participate, and to also decide about anthropometric and analytical measures according to the LOPD protocol.

Once this formality has been completed, the survey will be sent to be completed and to take anthropometric measurements, e.g., weight, abdominal circumference, height, etc., as well as biochemical levels, e.g., cholesterol, HDL, TG, glucose (Methodology Cobas). These measurements will be preferably taken at the time of the survey or at a later date agreed on with the patient. Both the survey and the anthropometric and biochemical measurements will always be respectively conducted and taken by the same dietician or medical staff member of the PO.

Patients will be informed about the group briefings to be given to those interested, and general guidelines about MS will be reported.

Training and advice. The objective of this training phase is to stir, if any, interest in knowledge of healthy habits, diet and physical activity. The information that will motivate the patient to practice healthy habits will seek to convey to the patient, so both understandable and rigorous time.

All the previously collected information will be systematised in appropriate files for computer processing and subsequent analyses by groups. This study will be a consistent part that will lead to conclusions of a statistical nature and will shape a "patient type." At the same time, we will provide a reference to be used in developing personalised dietary recommendations.

Briefings will last approximately 1 hour and will address groups of 6-7 people with similar characteristics: age/gender/have MS/healthy, but overweight/healthy and interested in healthy habits. The content of briefings will be presented as a PowerPoint presentation, and will be entertaining with illustrative graphs and images. Briefings will consist in a first general part for all groups, and a second specific part to address the characteristics of each group.

The general section will describe and justify the risk factors of suffering MS: genetic/ethnic predisposition, diet rich in saturated fats, sedentary lifestyle, changes in hormonal balance, etc. Following these introductory concepts, informative content, which will focus on nutritional treatment, will be offered. A complete, sufficient, varied, balanced and healthy diet can be key to successfully controlling obesity. The food pyramid will be presented by explaining the steps of each food type and how often it should be eaten.

3.4. Clinical

Personalised attention will be provided to those patients who show interest and have MS or are at risk of suffering it. Patients will be reminded to achieve the objective and will receive guidance, personalised recommendations, and a personalised diet will be provided. At 15, 30 and 45 days, patients will be followed-up to check their degree of compliance with diet and weight control, body composition, appetite and anxiety. A record of intakes will be completed during this period. After 60 days, the survey with the corresponding analytical tests will be passed to validate the dietary intervention.

The study will continue autonomously by patients over a 4-month period, during which they will not receive advice, but must apply the previously recommended guidelines. After this period, they will be given an appointment to go to the PO and to value their nutritional status.

The economic burden of this study lies mainly in the time spent on it by PO staff and the cost of material used: tools, biochemical strips, etc. If awareness is properly raised, financial support can be sought from laboratories in exchange for their participation as sponsors.

Moreover, there will be at point at which patients will be asked whether they will are willing to financially help with the study costs in the survey they complete in the clinical phase, which could ease the impact of the economic burden on the PO.

The human resources required are important because all the tasks require time and preparation, and the nature of the different tasks must be taken into account; easy communication to promote the project, rigour and consistency in devising surveys, and clinical qualification to determine biomarkers. While the project is underway, it has been estimated that two people will be necessary who will have a total load of average working hours.

4. EXPECTED RESULTS

Health systems now offer a wide range of treatments for chronic diseases to mitigate their consequences. The cost of many of these treatments amply justifies the benefits obtained in terms of the quality of life, but in some cases, financial limits hinder the implementation of such treatments. This is why any preventive intervention is essential because it helps reduce the healthcare costs caused by chronic diseases.

The proposed intervention performed from the PO, based on diagnosing MS, is expected to contribute to prevent MS in a large group of patients at risk of suffering it. Patients will be informed of the importance of MS and will be motivated for treatment in an attempt to improve their quality of life and to reduce social spending.

REFERENCES

[1] World Health Organization. Definition, diagnosis and classification of diabetes mellitus and Its complications. Report of a WHO consultation. Geneve: WHO; 1999.

[2] Zimmet, P. Alberti, G. Shaw, J. New World IDF definition of metabolic syndrome: arguments and results. *Diabetes Voice*, 2005; 50 (3): 31-33.

[3] Marju Orho-Melander. Metabolic syndrome: Lifestyle, genetics and ethnicity, *Diabetes Voice*, 2006; 51: 21-24.

[4] Mediterranean Diet Foundation. Mediterranean diet. Pyramid (accessed 9 April 2016). Available in: http://dietamediterranea.com/.

[5] Feinman RD, Volek JS. Carbohydrate restriction as the default treatment for type 2 diabetes and metabolic syndrome. *Scand Cardiovasc J*. 2008; 42: 256-63.

[6] Kallio P, et al. Dietary Carbohydrate you induce gene expression modification alteration in abdominal subcutaneous adipose tissue in in the metabolic syndrome With Personalities: the FUNGENUT Study. 2007 *Am J Clin Nutr*; 85: 1417-1427.

[7] Tsai, A. G., D. F. Williamson and H. A. Glick (2010), "Direct Medical Cost of Overweight and Obesity in the USA: A Quantitative Systematic Review" *Obesity Reviews*, 6 Jan., epub ahead of print.

[8] Withrow, D. and D. A. Alter (2010), "The Economic Burden of Obesit Worldwide: A Systematic Review of the Direct Costs of Obesity "*Obesity Reviews*, Jan. 27, epub ahead of print.).

[9] Popkin, B. M., S. Kim, E. R. Rusev, S. and C. Du Zizza (2006), "Measuring the Full Economic Costs of Diet, Physical Activity and Obesity-related Chronic Diseases" *Obesity Reviews*, Vol. 7, pp. 271-293.

[10] Foresight (2007), Tackling obesities: Future Choices, Project Report, Foresight, London.

[11] Matía Martin P, Pascual E, A. Pascual Nutrition and metabolic syndrome. *Spanish Journal of Public Health*. 2007; 81 (5).

[12] International Diabetes Federation. Final 1 Doc IDF Backgrounder: The IDF consensus worldwide definition of the metabolic syndrome. (Accessed March 5, 2016). Available in: http://www.idf.org/webdata/docs/IDF_Meta_def_final.pdf.

In: Modeling Human Behavior: Individuals and Organizations   ISBN: 978-1-53610-197-3
Editors: L. Jódar Sánchez, E. de la Poza Plaza et al.   © 2017 Nova Science Publishers, Inc.

Chapter 7

ACTORS AND FACTORS INVOLVED IN HEALTH TECHNOLOGY DIFFUSION AND ADOPTION: ECONOMIC, SOCIAL AND TECHNOLOGICAL DETERMINANTS

María Caballer-Tarazona, PhD and *Cristina Pardo-García, PhD*
Applied Economics Department, Universitat de València, Spain

ABSTRACT

Investment in health technology is a controversial issue because on the one hand, technological change is important for improving effectiveness of health care services, and on the other hand, new technology implementation it is considered one of the major drivers for rising costs.

Therefore, technology diffusion an adoption is a very complex mechanism which is affected for a combination of factors that are correlated. However, it is possible to identify and clustered the different factors which affect technology diffusion and adoption from different approaches.

Even if traditional theories predicted that new technologies will be adopted based on their expected cost and benefits; more recent research has identified social and organizational conditions as a relatively more important factor in technology adoption.

In this line, the aim of this chapter is to establish a common thread among reserches on the topic, clustering them in three groups in order to design a clear map of the variety of technology diffusion and adoption determinants. In addition, this review can be enlightening to cast the factors that encourage or impede health technology diffusion as well as the main issues that must be taken into account in the decision process of technology adoption.

Keywords: health technology, adoption, diffusion, economic and institutional factors, social factors, technology nature

---

*Corresponding author: María Caballer-Tarazona. Email: Maria.caballer@uv.es.

INTRODUCTION

The importance of technological change and technology adoption in the health care sector is an extensively discussed topic in economic literature.

Investment in health technology is indeed a controversial issue. Technological change is important for improving effectiveness of health care services, but new technology implementation is considered to be one of the major drivers for rising costs.

Economic literature in this field has focused particularly on the cost-effectiveness of large technologies due to their high unit cost which requires rigorous valuation studies. However, technologies with a lower unit cost may also produce a relevant impact on overall expenditures because of high volumes of purchase. Traditional theories have studied the process that leads to technology adoption based in particular on cost and benefits, but more recent research has adopted a wider perspective by including also social and institutional factors as determinants of technology diffusion.

Given this context, it is becoming increasingly evident that a better understanding of the factors that influence the diffusion of innovative technologies can improve models and rules of adoption, allowing more rational and effective technology diffusion.

The purpose of this chapter is to conduct a survey which highlights the main factors impacting adoption and diffusion of health technology. Literature that has directly addressed the issue of technology adoption and diffusion from a multifactor point of view is not extremely extensive in terms of number of contributions, but does cover a rather wide range of different methodological approaches. In particular, we identified three main streams of contributions that addressed the topic:

1. The first group includes studies where the problem of technology adoption and diffusion is considered in terms of economic and regulatory factors, and papers belonging to this area have a clear cost-containment orientation.
2. The approach of the second group is characterized by focusing on performance evaluation and technology availability in hospitals. In other words, the aim of the scholars is to identify and analyze the factors that emphasize or impede the adoption and diffusion, and consequently, the impact in the availability of new technology in hospitals, regardless economic factors. The main concern is not whether a particular technology displays features that make it worth being adopted, as technology diffusion is considered as a necessity to improve the quality of health services. Therefore, the focus of these contributions is on the social, organizational and networking factors which can affect the diffusion of health technology and which can be summarized in supply and demand factors.
3. The third group of studies considered here, comprises reserches that emphasize the role of the nature and specificity of the technology as determinant of its adoption and diffusion. Several studies have revealed differentiated diffusion patterns, depending on the specific characteristics of the innovation.

We often find in the literature studies of technology adoption and diffusion which take into account more than one approach and point out that technology diffusion depends on a combination of factors of different nature.

As proposed in this chapter, it is possible to cluster literature in the three groups based on the nature of the main factors studied and the approaches developed in each paper, in order to identify more clearly the variety of factors involved in the process of technology adoption.

METHODS

We surveyed here the main contributions appeared in the health economics literature concerning technology diffusion in order to critically assess the current state of art. Papers analyzed in this survey include studies characterized by mixed methodological approaches according to the criteria explained in the previous section. The unifying feature of the papers reviewed here is that they all address the same topic: the analysis of processes and factors in health technology adoption and diffusion, although this may occur with different methodological approaches and considering different sets of determinants.

We searched papers written in English, Italian and Spanish. As mentioned before, emerging technology diffusion is a complex phenomenon driven by a variety of factors and actors. Our claim is that despite such variety of contributions, identifying the underlying perspective that motivates different groups of works, and categorizing them accordingly, may shed new light on the results existing studies and may contribute to derive general and more comprehensive policy indications for the policy makers.

Technology adoption and diffusion is an issue influenced not only by economic factors but also by social and institutional factors. The analysis of each single factor can be addressed with different methodologies. Therefore, the different nature of these factors requiring consideration of different analytical methods, in order to enrich the vision of the problem, to allow take into account all the nuances from different perspectives, and consequently facilitate the understanding of the problem.

FINDINGS

"Cost Containment Orientation"

One of the main concerns regarding the adoption and diffusion of health technologies is its association with the observed growth in costs of health care. Rising medical expenditures is a political issue widely shared by developed countries, and innovation in medical technology has been regarded as one of its major drivers. Because of that, several studies focused on the economic and regulatory factors that may influence the dynamics of technology diffusion. In particular, institutional designs and reimbursement mechanisms which characterize different health care systems may influence technology diffusion. In this section we discuss the main points derived from a series of papers characterized by an approach focused on the necessity of establishing guidelines for cost containment. In this section, the effect on the diffusion process of different politico-economic systems was outlined.

Economic Factors and Reimbursement Models

From a macroeconomic point of view, the structure of the health care system can be identified as one of the factors that affect the different trends in technology diffusion. The organization of the health care system can promote or hinder the adoption and diffusion of new technologies. This diffusion varies across countries according, among other aspects, to the characteristics of regulatory policy and payment systems.

Considering this aspect, several papers analyzed the US health care system which has been traditionally characterized by rapid technological progress (Lettieri et al. 2009, Mas et al. 2008, Baker et al. 2001). In particular, these works analyzed whether the reduction in physician and hospital reimbursement associated to the growth in managed care, has slowed down adoption and has reduced the availability of technologies by limiting the use of expensive tests and procedures.

The impact of managed care on the hospitals' decision to adopt new technologies is theoretically not clear because the adoption decision is influenced by several correlated factors, therefore is not easy to identify the direct effect on technology adoption of managed care. In this section we try to assess what relationship emerges from the literature between managed care and patterns of health technology adoption and diffusion in the US market.

Traditional health insurers reimburse providers on a fee-for-service basis which barely controls utilization and allows insured patients to gain almost unlimited access to the providers of their choice. On the other hand, managed care policies place several restrictions on patients and utilization in order to reinforce the cost-containment. In addition, restrictions in the product choice offered to consumers are verified.

Price is a critical aspect for the selection of providers from a network. Managed care can affect adoption of technologies through different channels, because it has strong implications for the overall health care market by reducing medical care prices and affecting the physicians' practice patterns for other forms of health insurance as well. Hospitals not contracting with managed care organizations may be still influenced by managed care because they may be more inclined to adopt new technology in order to have access to a managed care network and avoid patients choosing other hospitals in order to take advantage of better price and services (Mas et al. 2008).

Baker et al. 2001 is one of the pioneering papers regarding the issues of managed care effect in technology diffusion. The aim of this paper is to examine through a hazard model the connection between the market share of HMOs (Health Maintenance Organizations) and the diffusion of magnetic resonance imaging (MRI) equipment in US hospitals between 1983 and 1993. Author studied the impact of managed care on MRI by comparing MRI diffusion and availability in markets with different levels of managed care activity. The model includes three sets of variables, one group of area controls; the second one includes determinants of health care demand and the third set for the characteristics of the area. Results indicate that increase in HMO activity is associated with slower diffusion of MRI equipment and lower overall MRI availability.

Later studies analyzed the topic with additional considerations in order to enrich the research. As it is described below, Mas 2008, considered in her study 13 technologies in order to build a model from which it is possible to obtain more general conclusions. The contribution of Bokhari 2009 instead is to distinguish effects on technology adoption for the cases of HMO penetration or HMOs competition.

Mas et al. 2008 use a hazard rate model in order to analyze whether higher levels of managed care market share are associated with a decrease in medical technology adoption during the period 1982-1995 in US hospitals. They introduce in the model different sorts of variables regarding insurance, hospitals controls, regulation and demographics. In addition, they created a unique data set with information on the cost reimbursement for each of the 13 technologies in order to evaluate to what extent managed care enrollment has larger negative effect on the adoption of less profitable technologies.

However, as studied by Bokhari 2009, the effect of managed care in technology diffusion can be different depending of the study perspective. For example, the point of Bokhari 2009 is to differentiate between HMO penetration and competition and examine their respective impact on the adoption of cardiac catheterization laboratories in US hospitals between 1985 and 1995. Authors estimated a hazard function under numerous specifications introducing in the model variables regarding HMO competition and penetration, population characteristics, area characteristics and hospital characteristics. Results show that hospitals are less likely to adopt the technology if HMO market penetration increases, but more likely to adopt it if HMO competition increases.

Therefore, within the managed care system it is possible to observe different trends of technology diffusion depending on the HMO penetration and competition. The HMO penetration decreases the probability of technology adoption by hospitals because a monopoly HMO can get big discounts from hospitals to see which ones reduced their profit and then slow down the spread of expensive technology. However, when the market increases the number of HMOs, and competition between them raise the likelihood of adoption of technology increases.

It must be noted, that this effect is not linear since it depends from neighboring hospitals having adopted the technology or not. Therefore, the diffusion of a health care technology is influenced by both the total market share of managed care organizations as well as the level of competition among them.

These contributions highlight the fact that managed care significantly affects the pattern of technology diffusion and adoption. In particular, managed care affects negatively patterns of adoption, slowing down diffusion of health technology (Mas et al. 2008, Baker 2001). However, this effect varies depending on other factors related to the specific technology. Results shows, in addition, that the effect of managed care is stronger for the less profitable technologies, in other words, the effect is stronger for technologies with higher cost-reimbursement ratios (CRR).

At the next level of analysis, it appears that reimbursement mechanisms are also important factors which affect the diffusion of technology, especially regarding new devices.

Prospective reimbursement systems based on DRGs (Diagnosis Related Groups) have become more and more popular in European countries in the last decades, due to the necessity to establish rational mechanism which allows a better understanding of the morbidity and costs related to a specific diagnostic.

After their implementation in US hospitals, DRG's became in the nineties one of the most remarkable applications of health financing in European hospitals [5]. Health technologies and devices can be reimbursed within global budgets assigned to hospitals on an annual basis or using the DRG tariff.

The DRG system ensures that professionals remain accountable of the gap between revenues and costs, because they can look at the specific DRG to understand the financial impact of their treatment decision. Institutional aspects such as choosing a DRG system induce a constant pressure on cost. When services are reimbursed prospectively according to DRG tariffs, professionals look at the specific DRG tariffs and the direct costs of the interventions for understanding the financial impact of their treatment decision, and this could discourage or encourage the adoption depending on the generosity of the remuneration.

Hence, when the reimbursement model is based on DRG, the relationship between tariffs and costs and the frequency of tariffs revision are two key elements for understanding the economic rationale behind the use of new technologies.

On the other hand, in a system reimbursed by global budget, the cost pressure is often softened by the possibility of renegotiation or bail out; therefore, those funding rules do not automatically translate into actual constraints. Consequently, with regard to funding technologies in this case, decision making is determined not only to a limited extent by cost-revenue considerations but it depends also on a combination of other factors as professional status or prestige.

Another group of papers which tackle the issue of technology adoption and diffusion by paying attention to the reimbursement model and other economic factors are Capellaro 2009, Vaughan 2010, Slade et al. 2001 and Lettieri et al. 2009. The goal of Capellaro et al. 2009 is to analyze coverage, procurement and reimbursement of three medical devices comparing the case of Italy and Spain. The research was carried out by reviewing published and grey literature, as well as national and regional legislation. In addition, 19 experts from hospitals and the industry were interviewed. The authors found that procurement and funding mechanisms can only partially explain organizational and professional behavior; the use of technologies is mainly left to professionals who are exposed to a variety of incentives.

As showed by the aforementioned paper which analyzes reimbursement methods for three medical devices (coronary stent, knee endoprothesis and defibrillator) in the Italian and Spanish public healthcare systems, it is possible to conclude that Spain may appear to be a more amenable setting for adopting these technologies because organizations have only "macro" constraints thanks to global budgets. In Italy however, DRG-based payments implies that professionals look at specific DRG tariffs and direct cost of the interventions, therefore remaining more accountable and influenced for the margin between revenues and cost.

Similarly, we can find other example of the reimbursement system effect for technology diffusion in the paper by Schereyogg et al. 2009. For the specific case of the introduction in 2002 of the novel DES (Drug Eluting Stent) system in Italy, even if the DES appeared to be an effective treatment option, the DRG associated with percutaneous trasluminal angioplasty (PTCA) did not draw any distinctions between different types of stents, thus providing no incentive for providers to use the costlier DES.

However, funding mechanism as isolated factor cannot explain the diffusion patterns. The conclusions show by Vaughan et al. 2010 and Slade et al. 2001 are also similar. Both authors, even if they apply different methodologies and approaches, conclude that economic factors as a determinant of technology diffusion are very conditioned by the nature of the technology. The main purpose of Vaughan et al. 2010 is to examine trends in the availability and use of coronary artery bypass graft (CABG) and percutaneous coronary intervention (PCI) in the US during the period of 1993-2004 in markets with and without Certificate of Need (CON) regulations for open-heart surgery or cardiac catheterization. This study concludes that even if more generous reimbursement will encourage the diffusion of a specific technology, there are other factors than can mitigate or reinforce their effect. For example, PCI is easier to implement than CABG, therefore the technology characteristics of PCI reinforced the effect of the reimbursement mechanism.

Other scholars addressed the analysis of economic factors which affect technology diffusion from a macroeconomic point of view. Slade et al. 2001 examine differences in the rate of diffusion of medical technologies in OECD countries between 1975 and 1995 by estimating equations for technology availability and utilization.

This paper uses data from 25 OECD countries from the period of 1975 to 1995. The study analyzes the growth of five procedures or technologies: MRI machines, CT scanners, liver transplant technology, kidney transplant technology and technology for hemodialysis patients. The purpose of the study is to analyze the relationship between per capita income and diffusion of the above mentioned procedures or technologies.

Results show that in general terms the income of countries is relevant only when explaining the timing of adoption of new technologies adoption in OECD countries, however is less important in explaining the long-term availability of these technologies. Therefore, higher income countries tend to be early adopters of new technologies, but the variation in availability is weakly related to cross-national differences in GDP per capita. The importance of income in explaining the long-term availability of a technology generally declines over time and becomes insignificant for some technologies.

In Lettire et al. 2009 the authors wanted to identify the relevant issues for technology assessment and selection at hospital level, and group them in a reference framework through an electronic search which collected the relevant contributions in the field.

Specifically, Lettieri et al. 2009 develops the concept of "value generation" regarding the technology adoption. They divided this value generation in social value, economic value and medical or technological knowledge. When discussing the economic value creation, the authors states how technology adoption increases revenues, along with improving the image and reputation. Applying a theoretical review methodology, it is possible to take a broader view of the problem and proposes a next step in which it is argued that the adoption of technology generates an economic value which in turn encourages the technology adoption and diffusion. Then, Lettieri et al. 2009 identify two main assessment perspectives: value generation and level of sustainability in the implementation stage. The two perspectives have been deployed in a list of 19 relevant issues that should be reviewed during the budget process.

We also can see in this example that factors which affect technology diffusion are correlated. Economic factors as the economic sustainability of the investment are significant elements affecting the adoption decision, but this effect could be strengthened or weakened by professional behavior, regulatory mechanisms and the nature of the technology.

When a budget committee in a hospital should consider the adoption of a certain technology, it is necessary to consider two issues: degree of self funding and ratio of fixed costs to variable cost during implementation stage. However, adopting a new technology requires also that new practices and routines must be institutionalized by health care professionals, because that professional behavior and conditions (technology acceptance among physicians, coherence to strategic goals, training intensity, coherence to the human physical resource and so on) could also affect the adoption.

Therefore, budget committee should assess the *organizational sustainability* of the investment. The assessment of a technology adoption should consider issues related to the external environment, such as coherence with the current legal framework and coherence with the generally accepted ethics and system values.

Summarizing this section, it seems obvious, that more generous reimbursement will encourage the diffusion of a specific technology, even if there are restrictive factors such as regulation. The reimbursement system implemented is also relevant. Reimbursement systems based on DRG could be more restrictive towards technology adoption because they link services and cost in a clearer way, therefore, this can disincentive the use of new and more expensive technologies or devices.

On the other hand, a reimbursement system based on general budget leaves more room for the election of more expensive technologies. In addition, the evidence suggests that the effects of reimbursement incentives are greater for purchases of diagnostic technologies than for lifesaving technologies. This proves the argument posed previously that economic factors could be conditioned by technology nature.

The economic factor must be considering in a holistic point of view during the adoption decision process. During this process, it is important to pay attention to some aspects such as organizational sustainability of the investment, the external environment or the technology life-cycle. In any case, economic factors alone cannot explain patterns of technology diffusion, because they are correlated with a complex variety of factors.

Regulation

The most direct strategy for health policy to slow down the rise in cost and therefore hinder the adoption and diffusion of health technology are regulatory mechanisms. Public health regulation reacts to industry innovation and developments, while public policies can impact on companies' behavior conditioning their production and prices through the adoption of defensive or favorable policies.

Strategies which emphasize the early adoption of innovative technologies produce a decreasing price and increasing utilization. Regional or national authorities which are interested in ensuring that citizens having timely access to a beneficial technology can establish a partnership with the manufacturers in order to achieve lower prices.

Depending on whether the regulation is focused on emphasizing cost containment (defensive strategy) or emphasized the early adoption (aggressive strategy), the effect on the technology diffusion will be different, not only among countries but also among regions within the same country. The case of the Certificate of Need (CON) in the US is a clear case of regulation conditioning health technology adoption. These regulations vary among the different US States, so the effect on the health technology adoption and diffusion varies among states.

A Certificate of Need (CON) is a state-administered regulatory mechanism designed to limit the supply of certain services by requiring hospitals to obtain approval before opening regulated services lines. Since 1984, 25 states had repealed CON laws entirely while others have repealed CON for certain services. However, the effect of CON regulation is, in turn, conditioned to other factors such as economic factors, opinion of physicians or technology nature.

In spite of this, regulation is an interesting issue which can affect technology diffusion both in a European and in a US context. Some papers which treat the issu are presented below.

The aim of the paper by Hashimoto et al. 2006 is to analyze how local practice norms and local clinical needs would affect the process of innovation diffusion. The paper provides a study case focus on two teaching high-tech hospitals in Japan and the US for the diffusion case of coronary stent. The study uses comparative data of both countries for the period between 1994 and 1998. This paper concludes that political regulation and economic incentive systems can just partially explain the process of technology diffusion. Other factors such as evolving nature of technology, the influence of local practice norms, and local needs shaped by patients' clinical characteristics, have a remarkable impact in the technology diffusion process.

In this line, the interesting study carried out by Vaughan et al. 2010, evaluates the impact of CON on the utilization of a specific technology in US hospitals. Presumably, compared with states without CON, states with CON may have fewer hospitals per capita performing CABG (Coronary Artery Bypass Grafting) or PCI (Percutaneous Coronary Intervention).

Results show that CON is effective at restricting the diffusion of PCI or CABG. In addition markets with significant repeal of CON had greater increases in CABG utilization. The impact of CON on cardiac catherization was different, with no evidence that repeal of CON was associated with growth in the absolute number or proportion of hospitals offering PCI or with growth in the utilization of PCI. That could be due probably to the nature of the technology, because PCI is relatively easy to learn and can be performed by a cardiologist in a catheterization lab, while CABG is a more complex operating room procedure requiring specialized surgical training. While the characteristics of PCI make it easy to implement and consequently encourage it spread.

Another factor that can weaken the effect of CON regulation in PCI interventions is the strong endorsement by clinical leaders. Therefore, even if CON is a regulatory mechanism that could inhibit the supply of certain technologies, as shown by the example given by Vaughan 2010 and Hashimoto 2006, the effect in the technology diffusion is conditioned by other factors of different nature. Specifically, the effect of CON regulation can be mitigated by the nature of the technology, economic factors and the behavior of clinical leaders.

As a matter of fact, in Baker 2001, Bokhari 2009 and Mas et al. 2008, the authors introduce in their model a control variable which considers states with CON as an indicator of states that have government regulation, and the results show that states with more severe regulation discourage technology adoption. However, even if the CON variable is statistically significant in the model, managed care has a stronger influence in technology diffusion, followed by the group of variables which defined hospital characteristics. All the papers reviewed in this section coincide with the idea that regulation can only partially explain the process of technology diffusion.

For European countries there seems to be evidence of a clearer effect of regulation mechanism in the patterns of technology diffusion. Regulation for technology adoption and diffusion has been applied mainly in the markets for devices in several European countries due to the dramatic increase in health expenditures. In the case of devices, as for the rest of health technologies, it is important that regulation finds the right balance between improving access to new medical devices and restricting market forces to contain cost and ensure affordability.

We found two studies point at the issue of regulation in a European context (Schreyogg et al. 2009 and Grilli et al. 2006). Both of these papers focus on the role of regulation for encouraging early adoption. The goal of Schreyogg et al. 2009 is to describe and discuss current policies for regulating devices in Europe. They explore the policies pursued by Germany, France, Italy and UK in order to find the right balance between improving access to new medical devices and restricting forces to contain costs.

Grilli et al. 2006, analyzes the introduction of DES in the Italian market, specifically in the Emilia Romagna region, and find clear evidence that the regulatory mechanism affects technology adoption and diffusion. This study case is quite significant because the Emilia Romagna region was pioneered in implanting some strategies in order encourage the early adoption of DES devices.

Shortly after DES was introduced to the Italian market, seven out of 19 regions adopted different programs to encourage their use. The Emilia-Romagna region elaborated a PTCA registry and developed guidelines in order to define people with a high risk of restenosis as the target-population for DES. The elaboration of guidelines which restricted the access to DES by clinical criteria, is an example of regulation in which regional government ensured that its citizens had access to beneficial medical devices but at the same time, establish a criteria of use in order to make it economically sustainable. In addition, the regional health authority negotiated with DES manufacturers in order to achieve lower prices for DES.

As shown in the different examples treated in this section, regulation can affect technology diffusion in different ways; there are strategies that encourage the early adoption of a technology and others which hinder it. In any case, even if institutional variables and regulatory structures can impact technology diffusion, their influence weakens over time (Bech et al. 2009).

Early Information

There has recently been increased interest in the provision of advice on emerging health technologies to decision makers. Availability of such advice can be useful to avoid uncontrolled adoption and diffusion of innovations that have not been properly assessed.

Early economic evaluation can serve as a useful tool supporting health policy and management for technology adoption decisions. In addition, an early evaluation of an innovation, may allow the design of more adjusted forms of reimbursement.

Nowadays, new technologies have to prove cost-effectiveness, affordability and benefits to the health care system before national health services or insurance schemes include them in the benefit package that is covered. It is difficult to determine the role that early economic data actually plays in decisions; in addition, empirical verification suggests that early economic data are not a standard tool in public policy decision-making (Hartz et al. 2009). Sometimes it is advisable to provide prompt access to an effective technology even before the evidence is available. Due to the acceleration of technological progress, the expiration of the assessments is shortened. However, delaying the adoption decision increases rapidly the knowledge gap.

In US, due to this reason, less than half of medical care is based on strong and robust evidence of effectiveness (Gónzalez López-Valcárcel 2007). Then, new technologies are often made available at an "immature" stage of the development. This is the case for medical devices, whose current regulation in Europe does not require to the industry to establish efficacy prior to their market launch.

An interesting example is the case of Emilia-Romagna described before; where the DES was introduced before the availability of clinical data in order to ensure that its citizens had timely access to the beneficial medical device (Grilli et al. 2006). Nevertheless, the fact that the new stent was used for conditions never tested in clinical trials calls into question the extent to which innovation may have an impact on quality of care. As shown the experience in Emilia-Romagna, a regional registry helped monitor practice, providing systematic assessment of the cost-effectiveness of the new stent and its impact on alternative surgical treatments. This register allows policy-makers the opportunity to monitor diffusion and possibly reverse the initial decision.

In Spain, the introduction of new devices is a processed quite formalized. Clinicians are required to provide the scientific evidences of therapeutic value in order to introduce a new technology. This evidence is evaluated by an internal commission that includes managers and clinicians. Therefore, in this case, the availability of scientific evidence regarding a specific technology can contribute to the decision-making processes, and consequently, can affect the technology diffusion trend (Cappellaro et al. 2009).

Availability of prompt advice regarding impacts of the introduction of a new technology is also studied by Hartz et al. 2009 and Hailey et al. 2001. The aim of the paper by Hartz et al. 2009 is to explore the different ways in which early economic data can inform public health policy decisions on new medical technologies. This paper conducted a literature research to detect papers addressing to identify contributions of early economic assessments as well as economic evaluations that actually used data from early phases of product development. The results show that decision-makers can benefit from the information supplied by early economic data. However, it is difficult to determine the role that early economic data actually plays in decisions due to that empirical evidence in this field is not wide.

Decisions regarding health policies related to emerging innovative technologies are often rather transparent. Hailey et al. 2001 elaborated a pilot project to provide advice to decision makers on new and emerging medical technologies. Within this pilot project briefs on technologies which were not yet available and which might have a significant impact on health care were prepared.

Given the fact that health care policy makers have usually little time and opportunity to read technical reports, it seemed desirable to provide advice in a concise form. The project developed in this paper underlyed some issues for consideration in the process of providing advice:

- Process details: Effective scanning for relevant emerging health technologies requires familiarity with the local and national health care system, and experience in health technology assessment.
- Implications for the policy process: While there were indications that some briefs had been helpful to decision makers and responses to the survey were positive, there were a mismatch between the rapid preparation and delivery of the briefs and the slower and less focused processes within the health policy-making agencies.

The experience of the project described in this paper confirmed the value of providing short alerts on emerging health technologies for a provincial health care system and in general, the value of expanding the HTA (Health Technology Assessment) knowledge base.

Papers reviewed in this section show that even if early information can be very useful in health policy design and during the technology adoption decisions, this information is not always used appropriately due to lack of transparency in the process, long bureaucratic processes and in general, lack of structures capable to manage, assess and apply this kind of information to the technology adoption decision process.

"Achievement and Performance Orientation"

A second body of studies includes papers which address the topic of health technology diffusion with an approach focused on performance and achievement. This stream of literature emphasizes the basic purpose that lies behind technology innovation, which is improvement of health care, achieving higher standards in the quality of health services, with relatively low attention to the economic or budget implications.

Within this group of contributions, it is possibly to divide the factors which affect technology adoption and diffusion in demand factor and supply factors.

Demand Factors

Traditional theories predicted that new technologies will be adopted based on their expected cost and benefits; however, more recent research in economics has identified social conditions as a relatively more important factor in technology adoption. In fact, several scholars consider physicians and networks as a very important variable in the adoption of new technology (e.g., Bo Poulsen et al. 2001, Jippes et al. 2010 and Capellaro et al. 2009).

Physicians, as expression of the demand side of the market, behave according to their own preferences, their experiences (learning by using) and arising network externalities. All these factors contribute to the decision to adopt a certain technology. The knowledge about the existence of a new medical technology, which involves a more effective medical treatment, is diffused by interacting physicians who make the decision to adopt a new technology.

As mentioned before, many studies consider the role of physicians and network as an important factor in the adoption and use of new technology. In the literature we found several examples of papers that treat this issue: Jippes et al. 2010, Burke et al. 2007, Burke et al. 2009, Grebel et al. 2010, Fitzgerald et al. 2002 and Cappellaro et al. 2009. In Jippes et al. 2010 the authors examine the effect that following an intensive Teach-the-Teacher training had on the dissemination of a new structured competency-based feedback technique of assessing clinical competencies among medical specialist in the Netherlands. The results show no effect for Teach-the-Teacher training course on the dissemination of the new structured feedback technique and a strong effect for network tie strength. The results also show a negative effect of physician's age on new technology use; therefore, younger physicians are more likely to use new technologies. This paper finds that the effect of networks has a stronger effect on new technology diffusion than formal trainings.

Similar to this, also Grebel et al. 2010 focus their study in the role of networks and interactions among physicians on the technology diffusion process. Concretely, Grebel et al. 2010 investigates the diffusion process of two competing innovative technologies in the health care sector. Specifically, the paper treats the case of percutaneous aortic valve replacement. The demand side is modeled using a social learning approach, where the knowledge about the existence of a new medical technology, which involves a more efficient medical treatment, is diffused by interacting physicians who make the decision to adopt a certain design of a new technology.

As in other fields, the process of knowledge creation does not end with the first adoption of new technology. In medicine, the users accumulate experience and know-how by using the technology and this experience could deliver a feedback for other physicians and for the technology industry. Learning by using thus contributes to a continuous improvement of a new technology or, in the negative case, may lead to abandoning it.

With an increasing number of users, direct network externalities occur and the increasing importance of a new technology may additionally stimulate the innovation of complementary products. These reinforcement effects accelerate the technology's diffusion.

The role of the opinion leaders within these networks has also generated interest in the literature. In two related papers by Burke et al. 2007 and Burke et al. 2009, the diffusion of coronary stent in the presence of prominent physicians (stars physicians) within a local peer group was studied within Florida hospitals.

The aim of the paper is to analyze the social influence among physicians in the adoption timing and utilization of coronary stent. The starting hypothesis is that "start" physicians have a positive and strong effect on the adoption timing and utilization of coronary stent by non-star physician. "Star" status is defined as having completed residency at a top-ranked hospital. Results show that the diffusion of stents by non-stars depends positively on the number of stars practicing contemporaneously at the same hospital. However, social influence in the opposite direction it was not found.

In addition, findings indicate that the lack of star physicians may slow adoption. In other words, the opinions of "star" physicians influence the opinion of other physicians, because given their expertise, they may instruct others, either directly or by example, in the proper execution and application of the procedure; such instruction may be more effective than having each individual learn from primary sources. It could be that star physicians have superior ability to integrate the results of research studies and engage in informal communications.

In the second study by Burke et al. 2009, the point of the authors it is to experiment with an alternative construction of star status and with additional control variables in the analysis. Star status definition was also restricted as having completed residency at a top-ranked hospital in the last 10 years. New results strengthen the conclusions of the previous paper, in addition was found that the social influence of star physician in timing adoption and utilization is stronger when the definition of star status is restricted to those with more recent residency training. In other words, younger "star" physicians have a stronger effect on coronary stent diffusion. Therefore, the age of physicians is another factor which could negatively affect technology diffusion.

A less recent paper also treats the issue of opinion leaders and networks in technology adoption and diffusion. Fitzgerald et al. 2002 take as case study two researches on the diffusion of innovation in acute and primary care sectors in the UK. The authors analyze different case study in order to compare the role of network in different context, which allowed examining the trajectory of the innovations in order to explore formal and informal processes and to examine the views of a range of different stakeholders.

Authors consider formal processes those such as formal training courses, and informal processes those such as direct instructions or examples given by physicians' leaders regarding the use of a new technology.

This study demonstrates that processes of diffusion are deeply affected by interprofessional relationships in each context. Networks can engage people in the diffusion process or they can halt the process. Often, relationships of trust and respect were able to counterbalance negative contextual factors.

Therefore, it was proved in several studies that opinion leaders play an active and influential role in the diffusion of innovation. In addition, Fitzgerald et al. 2002 identifies three roles that can be played by the opinion leader. 1. A node or focal point for information, who may act as a link between academic research and practice. 2. An expert opinion leader, with local creativity. 3. A strategic political opinion leader, with combined management and political skills.

Each of these roles may facilitate or inhibit diffusion. On the other hand, within the inner context of the organization's boundary, the history, culture and quality of inter professional relationships will be factors which account for variation in rate of diffusion.

As illustrated in this section, evidence shows that professional networks have a positively effect on the dissemination of new techniques (Burke et al. 2007, Burke at. al 2009). Physicians with strong ties are more likely to show adoptive behavior (Jippes et al. 2010 and Bo Poulsen et al. 2001), and this effect is reinforced for younger physicians.

Even if the reimbursement systems, economic and regulation issues can influence professional behavior, factors such as professional values, scientific prestige and reputation effects may motivate physicians to act in contrast to the financial incentives.

Consequently, physicians could be considered one of the key factors for technology diffusion. Networking emerges as more important factor than physician background or education. Physicians with a strong tie to network are more likely to adopt new technology (Jippes et al. 2010 and Fitzgerald et al. 2002). Training variables as short training courses have not any effect on the adoptive behavior, or at least are not influential enough for adopting innovations successfully.

Other physician characteristic which affects the innovative behavior is age. As concluded in studies such as Jippes et al. 2010 or Burke et al. 2009, age has a significantly negative relationship to adoptive behavior; in other words, young physicians are more likely to have an adoptive behavior.

Supply Factors

In the supply side of health care services, the focus on the factors which affect technology adoption and diffusion is shifted towards hospital characteristics, in particular the overall size of the hospital. This influence is analized in Baker 2001, Nystrom et al. 2002 and Bo Poulsen et al. 2001.

Nystrom et al. 2002 explore the role of organizational climate and organizational context on innovation in hospitals of US. Context refers to organizational size, slack resources, and organizational age. The paper examines a geographically homogeneous sample of a targeted population in hospital industry so that they can control for other factors that may influence technological innovation. Results show that organizational size and slack are positively related with conservativeness. Hierarchical regression analyses indicate that the climate measures of risk orientation and external orientation interact significantly with the context dimensions of organizational size and organizational age.

Secondly, Bo Poulsen et al. 2001 conclude that hospital size positively affects technology diffusion, but it doesn't always represent a significant variable regarding time adoption. Concretely, Bo Poulsen et al. 2001 analyze the impact of different hospital characteristics on the hospital adoption of LC (Laparoscopic Cholecystectomy) in Denmark and in Netherlands.

Even if the patterns of diffusion are quite similar in both countries, some differences were found. The results show that the size of the hospital had a positive influence upon the timing of adoption of LC in Denmark, but not in the Netherlands. Characteristics such as location of the hospitals or teaching status did not influence the timing adoption of LC. According to the literature, the finding of this study may be extrapolated to an international perspective in order to suggest that hospital size plays a prominent role in the diffusion of LC.

In a previous paper, Bo Poulsen et al. 1998, the authors analyzed the diffusion of five laparoscopic technologies in Denmark hospitals and concluded that large and specialized hospitals were the earliest adopters.

Nystrom 2002, Bo Poulsen 2001 and Bo Poulsen 1998 pay attention to the diffusion of the following technologies: specifically magnetic resonance imaging, imaging technologies and laparoscopic technologies respectively. All of them concluded that larger hospitals are likely to adopt technology and positively influence the time of adoption. Organizational size directly and positively affects innovations. Local conditions can also influence the role of hospital size. As an example, according to Bo Poulsen 2001, in Holland hospital, size is not a significant characteristic in the technology diffusion of LC, perhaps because the Dutch hospitals are on average big enough to adopt this technology, so there is not a variable capable to discriminate the time of adoption.

The status of the hospital also affects technology diffusion. Teaching hospitals or specialized hospitals are more likely to adopt new technology. New applications are expected to be found first in specialized hospitals, because of their innovative behavior. As concluded by Baker 2001, teaching hospitals or more specialized hospitals are much more likely to adopt technology. Bo Poulsen et al. 1998 also conclude that hospitals characterized as specialist in training are associated with an earlier adoption of technology (LC).

The location of the hospital could also affect the adoption of a specific technology, because as concluded by Baker 2001, the adoption of a technology by a hospital depends on the neighborhood. If others hospitals in the area have already adopted the technology, the probability of adoption decrease. However, if by "location" we mean the fact of a hospital being rural or urban, the evidence concludes that location is a factor which has no significant effect in technology adoption (Kempt et al. 2008).

"Technology Nature"

As mentioned along all this review, health technology diffusion is determined by a complex and varied group of factors. Variations in the process of technology adoption and diffusion could be explained not only by political regulations, economic incentive systems or demand and supply factors, but also by the evolving nature of the technologies. In general, a technology should be more likely to have a fast diffusion if it is easy to use, if there is clear evidence that it performs better than the technology it substitutes, and if it is not used for emergency procedures.

In the literature we found some studies which point at technology nature as one of the main factors which affect technology diffusion. Firstly, Kemp et al. (2008) study the effect of rural versus urban hospitals in the laparoscopic diffusion. The scholars present a descriptive comparison of the adoption rate of laparoscopic cholecystectomy in small rural versus urban hospitals in the US. The aim of the paper is to check if professional isolation of rural physicians serves as an obstacle to the adoption of new techniques. The results show that most rural surgeons successfully overcame professional isolation in learning and adopting laparoscopic cholecystectomy. The authors concluded that even if hospital location has no significant effect on the diffusion of this technology, the nature of the technology it can conditioned its diffusion.

Therefore, for a procedure such a laparoscopic cholecystectomy there was no delay in the adoption of new surgical techniques or decline in surgical quality because of professional isolation. However, there was a delay in using the newer techniques for more urgent procedures. Urgent cases can be more complicated and technically challenging compared with elective cases.

Bo Poulsen et al. 1998 use a different methodology to reach similar conclusions than Kemp et al. 2008. This paper investigates the determinants of the diffusion of five laparoscopic technologies in Denmark. Questionnaires on 17 potentially influential factors on the adoption were sent to 59 hospitals. Results show that factors such as nature of the technology, training, competition and media attention have stimulated the diffusion, whereas budget for investment, budget for operation and public regulation usually had an impeding effect. According with this study, one of the main factors which influence the adoption is the nature of the technology. For the case of laparoscopic technology, even if the diffusion of laparoscopic cholecystectomy has been wide and fast, the adoption of laparoscopic appendicectomy has not had the same diffusion. The technology was also used infrequently by the adopting departments. Reasons might be that appendicitis is an emergency condition and on the other hand there is no evidence to justify substituting laparoscopic for the standard appendicectomy.

Finally, as shown in a previous section, Vaughan et al. 2010 studies the impact of CON on the utilization of CABG and PCI in US hospitals, and concludes that the effect of CON regulation could be mitigated by the nature of the technology. Specifically, this study argued that the characteristics of PCI, which makes it easy to implement, are one of the main factors which mitigated the effect of CON regulation in the PCI diffusion.

All papers in this section coincided in state that technology nature affects significantly its adoption and diffusion. As hypothesized at the beginning of the section, when a technology is easy to implement or there is clear evidence that it performs better than the technology it substitutes, faster trends of adoption and diffusion are more likely to appear, even if there are economic or regulation factors which could be restrictive for the technology diffusion. In addition, technology which is easy to use and applicable to large numbers of patients is most likely to be adopted even without evidence (Shih et al. 2008).

Furthermore, the literature agrees to conclude that health technologies used for emergency procedures has a slower diffusion compared to technologies related to elective services.

DISCUSSION

The decision to adopt a new health technology should be based not only on its effectiveness but also on local conditions such as local needs and norms. Health policy researchers should regard the process of technology adoption and diffusion as a dynamic process affected by patients, physicians, hospitals and technologies characteristics.

It is important to accept that there can be no uniform pattern in the diffusion of innovations. Adoption and diffusion will be influenced by interplay of factors: the credibility of the evidence, the characteristics of the multiple groups of actors, of the organization itself, and of the characteristics of the outer and inner contexts.

Even if technology diffusion is a very complex mechanism which is affected for a combination of factors that are correlated, it is possible to identify and cluster the different factors which affect technology diffusion from different approaches.

This paper tries to unify the literature regarding technology diffusion and adoptions, taking into account the different methodologies and approaches that tackle the topic. In order to clarify the map of the variety of factors which affect technology diffusion, the literature reviewed has been clustered in three groups. The first one embraces those papers which put special emphasis in economic and institutional factors. The second one includes papers that take into account social factors, both from the demand and the supply side. Finally, the third group of papers focuses on the effect of technology nature in its diffusion. This classification allows us to draw some conclusions and clarifications.

As far as the structure of health care system is concerned, the literature shows that for the case of US, the evidence indicates that managed care has a negative effect in technology diffusion. However, this effect could be different if we consider HMO penetration or competition. HMO penetration decreases the probability of technology adoption while HMOs competition increases the likelihood of adoption.

Regarding the economic factor, it is obvious that more generous reimbursement mechanism can incentive the adoption and use of a new technology. In addition, literature shows that a reimbursement system based on a general budget should be laxer reporting the margin between revenues and cost. On the other hand, in a reimbursement system based on DRG system, the link between service and cost becomes clearer and consequently, it could disincentive the use of a new costlier device.

From a macroeconomic point of view, the literature shows that the income of a country is significant when explaining the early adoption, but not the long-term availability of medical technology.

As far as regulation, it can affect technology diffusion in different ways; there are strategies that encourage the early adoption of a technology and others which hinder it.

With regards to early economic data, it can serve as a useful tool supporting health policy and management for technology adoption decisions, among other reasons because from an economic point of view, the control of technology diffusion should be judged in the light of efficient resource utilization. However, availability of early information about a new technology is not always used appropriately due to several elements as for example, the lack of transparency in the decision process of technology adoption, long bureaucratic processes that difficult the implementation of the technology, and in general, lack of structures capable to manage, assess and apply this kind of information to the technology adoption decision.

In the demand side, the role of opinion leaders and networks has been identified as a key factor in the process of new technology diffusion. Approaches which consider social factors as crucial in the technology adoption and diffusion process are relatively recent. In fact, the traditional studies on this field are generally based on economic issues and, in particular, on a cost profit approach. Recent evidence shows however that professional networks have a positive effect on the dissemination of new techniques. Physicians with strong ties are more likely to show adoptive behavior, and this effect is reinforced in younger physicians.

As far as the supply side is concerned, most authors agree to conclude that variables such as hospital size are of the more significant one in explaining the trends in technology diffusion. Larger hospitals are more likely to adopt new technologies. In some cases, the status of specialization of a hospital can also be a positive element in the adoption process.

Finally, all the factors identified along this literature review are strongly conditioned by the technology nature. The literature shows how technologies which are easier to implement can mitigate the negative effects of economic or regulation variables on the diffusion. It appears that the technologies aimed at emergency departments are more reluctant to the introduction of novelties.

It is evident, that the process of health technology adoption and diffusion is complex because it depends of the interaction of numerous factors and actors with different interests. It would be therefore advisable that government agencies, industry and physicians' groups work together to determine the balance ensure that patients have access to novel, potentially lifesaving technology and at the same time to keep a moderate cost raising (Shih at al. 2008).

In any case, the identification and classification of the various factors involved in the adoption and diffusion process can be useful for the design of dissemination strategies for a particular technology, or in any case, for drawing a complete map of the health technology adoption and diffusion issue as a base for ulterior empirical studies in this field.

REFERENCES

Baker L. (2001). Managed care and technology adoption in health care: evidence from magnetic resonance imaging. *Journal of Health Economics*; 20:395-421.

Bech, M et al. (2009). The influence of economic incentives and regulatory factors on the adoption of treatment technologies: A case study of technologies used to treat heart attacks. *Health Economics*. 18:1114-1132.

Bo Poulsen P., Adamsen S., Vondeling H., Jørgensen T. (1998). Diffusion of laparoscopic technologies in Denmark. *Health Policy*; 45: 149-167.

Bo Poulsen P., Vondeling H., Dirksen C., Adamsen S., Go P., Ament A (2001). Timing of adoption of laparoscopic cholecystectomy in Denmark and The Netherlands: a comparative study. *Health Policy*. 55: 85-95.

Bokhari F. (2009). Managed care competition and the adoption of hospital technology: The case of cardiac catherization. *International Journal of Industrial Organization*; 27: 223-237.

Burke M., Fournier G., Prasad K. (2009). The diffusion of a medical innovation: Is success in the stars? Further evidences. *Southern Economic Journal*; 75(4): 1274-1278.

Burke M., Fournier G., Prasad K. (2007). The diffusion of a medical innovation: Is success in the stars?. *Southern Economic Journal*; 73(3): 588-603.

Cappellaro G., Fattore G., (2009). Torbica A. Funding health technologies in decentralized systems: A comparison between Italy and Spain. *Health Policy*; 92:313-321.

Fitzgerald L., Ferlie E., Wood M., Hawkins C. (2002). Interlocking interactions, the diffusion of innovations in health care. *Human Relations*. Volume 55(12): 1429-1449.

González López-Valcárcel B. (2007). La incorporación de nuevas tecnologías en el Sistema Nacional de Salud. Coste-efectividad y presiones sobre el gasto sanitario [The incorporation of new technologies in the National Health System. Cost-effectiveness and pressures on health spending]. *Presupuesto y Gasto Público* [Budget and Public Expenditure]. 87-105.

Grebel T., Wilfer T. (2010). Innovative cardiological technologies: a model of technology adoption, diffusion and competition. *Economics of Innovation and New Technology*. Vol. 19, N. 4: 325-347.

Grilli R., Taroni F. (2006). Managing the introduction of expensive medical procedures: use of a registry. *Journal of Health Services Research and Policy*. Vol. 11. N. 2: 89-93.

Hailey D., Topfer L., Wills F. (2001). Providing information on emerging health technology to provincial decision makers: a pilot project. *Health Policy*. 15-26.

Hartz S., John J. (2009). Public health policy decisions on medical innovations: What role can early economic evaluation play?. *Health Policy*; 89: 184-192.

Hashimoto H., Noguchi H., Heidenreich P., Saynina O., Moreland A., Miyazaky S., Ikeda S., Kaneko Y., Ikegami N. (2006). The diffusion of medical technology, local conditions, and technology re-invention: A comparative case study on coronary stenting. *Health Policy*; 79:221-230.

Inoriza I., Coderech J., Carreras M., Vall-llosera L., García-Goñi M., Lisbona J., Ibern P. (2009). La medida de la morbilidad atendida en una organización sanitaria integrada [The measure of morbidity in an integrated health organization]. *Gaceta Sanitaria*; 23: 29-37.

Jippes E., Achterkamp M., Brand P., Kiewiet D., Pols J., van Engelen J. (2010). Disseminating educational innovation in health care practice: Training versus social networks. *Social Science and Medicine*; 70: 1509-1517.

Kemp J., Zuckerman R., Finlayson S. (2008). Trends in adoption of laparoscopic cholecystectomy in rural versus urban hospitals. *American College of Surgerons*. Vol. 206, N. 1.

Lettieri E., Masella C. (2009). Priority setting for technology adoption at a hospital level: Relevant issues from the literature. *Health Policy*; 90: 81-88.

Mas N., Seinfeld J. (2008). Is managed care restraining the adoption of technology by hospitals?. *Journal of Health Economics*; 27: 1026-1045.

Nystrom P., Ramamurthy K., Wilson A. (2002) Organizational context, climate and innovativeness: adoption of imaging technology. *Journal of Engineering and technology management*; 19: 221-247.

Schreyögg J., Bäumler M., Busse R. (2009). Balancing adoption and affordability of medical devices in Europe. *Health Policy*; 92: 218-224.

Shih Ch et al. (2008) Diffusion of new technology and payment policies: Coronary Stent. *Health Affairs*. 27, n. 6: 1566-15.

Slade E., Anderson G. (2001). The relationship between per capita income and diffusion of medical technologies. *Health Policy*; 58:1-14.

Vaughan M., Bayman L., Cram P. (2010). Trends during 1993-2004 in the Availability and use of revascularization after acute myocardial infarction in markets affected by Certificated of Need Regulations. *Medical care Research and Review*. Volume 67. Number 2. 213-231.

Wilson, Ch. (2006). Adoption of new surgical technology. *BMJ*. Vol; 332:112.

Chapter 8

ASSESSING THE OPERATION OF AND USER SATISFACTION WITH THE ELECTRONIC PRESCRIBING SYSTEM IN THE VALENCIAN COMMUNITY (SPAIN)

Isabel Barrachina[1,*], *Elena de la Poza Plaza*[1,1],
Beatriz Pedrós[2,2] *and David Vivas*[1,3]

[1]Centro de Investigación en Economía y Gestión de la Salud, Facultad de Administración y Dirección de Empresas, Universitat Politècnica de València, Valencia, Spain
[2]Oficina de Programas Farmacéuticos, Subdirección General de Posicionamiento Terapéutico y Farmacoeconomía, Dirección General de Farmacia y Productos Sanitarios, Agencia Valenciana de Salud, Spain

ABSTRACT

The objective of the present chapter is to assess the operation of and user satisfaction with the electronic prescribing system (known as RELE) and to identify any aspects that could improve it. To do this, a questionnaire was validated and sent to a sample of healthcare users (n = 587) in the province of Castellón. The structured questionnaire included these sections: 1) demographic data; 2) prescriptions; 3) dispensing; 4) adhesion to and complying with treatment; 5) overall satisfaction. After the descriptive statistical analysis of the obtained responses, logit and CHAID analyses were done to know patients' degree of satisfaction. The obtained results revealed that 81.9% of those surveyed considered that the RELE system offered them advantages as health system users; 60% stated they visited their medical centre less since this electronic prescribing system came into being. The aspects that more strongly influenced the satisfaction of

---

[*] Isabel Barrachina: Centro de Investigación en Economía y Gestión de la Salud Facultad de Administración y Dirección de Empresas. Edificio 7J. Universitat Politècnica de València Camino de Vera s/n 46022 Valencia, Spain. Email: ibarrach@esp.upv.es.
[1] Elena De la Poza e-mail: elpopla@esp.upv.es.
[2] Beatriz Pedrós e-mail: pedros_bea@gva.es.
[3] David Vivas e-mail: dvivas@upvnet.upv.es.

patients with chronic diseases with the RELE system were: having to go to their medical centre less frequently (OR ratio 2.413), the quality of the information on treatment sheets (OR ratio 3.646) and the time spent with one's doctor not being cut (OR ratio 3.352). To conclude, the RELE system has improved the prescribing and drug dispensation process, reduced the frequency patients visit their medical centre, enhanced accessibility and helped dispensing in their chemists.

Keywords: eletronic prescribing system, Valencian Community, chemists, user, quality, satisfaction

INTRODUCTION

The electronic prescribing system coordinately combines a prescribing and dispensing system by means of a computer system, favours the rational use of medication, and guarantees patient safety by cutting down on prescribing and dispensing errors [1-7]. It allows complete therapeutic plans to be devised for patients, which is a particularly advantage for patients with chronic diseases who are on the same medication for a long time [8]. Administrative proceedings are reduced, which implies better access [6, 9, 10] to the public health system and better mobility around Spain, provided that all the electronic prescribing systems of all the Spanish Autonomous Communities (SACs) are included [11]. It also enables an alert system to be set up [12] to help prevent adverse medication episodes.

The main objectives of the electronic prescribing system are: reducing the medical errors that occur when wrongly administering medicines; encouraging a more rational use of medications; avoiding medications being stored unnecessarily; making the prescription dispensing system in chemists' quicker; cutting the number of chronic patients' visits to their medical centres. All these objectives help cut healthcare costs.

In 2005, the electronic prescribing system began to be set up in Spain [13, 14]. In 2001, the Valencian Health Agency (AVS) of the Valencian Community (VC) started the adaptation work needed to set up this electronic system, known as RELE [15-17]. This work involved organisational, legal and contractual changes [18-20]. However, it was not until March 2008 when RELE began to be set up in the province of Castellón.

This study assesses the present RELE system compared to the former system for the period from January 2008 to December 2009. As users remembered the previous system, they were able to evaluate the newly established system. This assessment was made by means of a questionnaire, which was arranged in the doctor's prescribing and the chemists dispensing stages. It also assessed user satisfaction with the new system.

The compulsory aspects that the RELE system must fulfil were assessed: a) registering the magnetic reader on the patient's medical card in both the prescribing and dispensing processes, which is always returned to patients; b) administering therapeuric treatment sheets to patients; c) administering a receipt of dispensed medications (Law 29/2006 on Guarantees and the Rational Use of Medications and Health Products) [21].

METHODS

An observational cross-section study was conducted using a structured questionnaire about how the RELE system operated and the degree of patient satisfaction. The study area included medical centre users in departments 1, 2, 3, 4 and 11 of the VC, where the assigned population included 441,526 inhabitants. The questionnaire was devised by a panel of experts (who were responsible for implementing the RELE system). The questionnaire contained 48 questions arranged into six dimensions to cover all the objectives set out:

- Survey taker's data and the questionnaire's identification data: date, department, centre and the patient who answered it (6 questions).
- Informant's classification data (8 questions).
- Prescribing system (10 questions).
- Dispensing system (18 questions).
- Therapeutic use, adhesion and compliance (2 questions).
- Aspects about overall patient satisfaction (4 questions).

After an initial pilot trial with 90 questionnaires, the definitive questionnaire version was written, which was filled in by interviewing patients directly in November and December 2009. A sample of 587 questionnaires from 31 medical centres was formed (sampling error <3.5%). Questionnaires were assigned to each medical centre in a direct proportion to the number of patients per centre by rounding off to the nearest whole number (Table 1).

To analyse how the RELE system worked, the relative and absolute frequencies of each obtained response were calculated. The CHAID analysis [22] ("Chi Square Automatic Interaction Detection") was applied with the DYANE-3 programme to study the relation between patient satisfaction with the system and the variables introduced into the study to sequentially define the explanatory variables that most strongly influenced the variable satisfaction.

By logistic regression, the joint influence of the explanatory variables (independent variables (independent, regressors or covariables) was quantified, considered predictors of the probability (P) of user satisfaction with the RELE system. This was done with the SPSS statistical programme.

In general terms, the mathematical expression of the binomial logistic regression is:

$$Ln\left\{\frac{P(y=1/x_1,x_2,....x_n)}{P(y=0/x_1,x_2,...x_n)}\right\} = \alpha + \beta_1 * x_1 + \beta_2 * x_2 + ... + \beta_n * x_n \quad (1)$$

where Y is a binary variable. It took a value of 1 when the patient was satisfied with the RELE system, and 0 otherwise.

The OR quotient indicated how many times more likely the user felt satisfied with the RELE system compared to if he/she was not satisfied (probability of success *vs.* failure). This ratio is defined as:

$$Odds = \frac{P(y=1/x_1,x_2,......x_n)}{P(y=0/x_1,x_2,...x_n)} = \frac{P(y=1/x_1,x_2,...x_n)}{1-P(y=1/x_1,x_2,......x_n)} = e^{\alpha + \beta_1 * x_1 + \beta_2 * x_2 + ... + \beta_n * x_n} \quad (2)$$

Table 1. Distribution of questionnaires per medical centre with the assigned population and quota percentage of the total

MC	CENTRE	ZB	QUOTA	% QUOTA	N	N'
Department 1: Vinaroz						
29	MC Vinaròs	7	28,357	6.42%	37.71	38
Department 2: Castellón						
37	MC L'Alcora	3	12,041	2.73%	16.01	16
40	MC Almazora/Almassora	4	20,898	4.73%	27.79	28
46	MC Auxiliar Oropesa del Mar	5	7,639	1.73%	10.16	10
54	MC Castellón (Sant Agustí)	7	6,180	1.40%	8.22	8
56	P Castellón (Castalia)	7	19,733	4.47%	26.24	26
57	MC Castellón (Pintor Sorolla)	8	13,371	3.03%	17.78	18
202	MC Castellón (Gran Vía)	9	10,427	2.36%	13.87	14
60	P Castellón (9 d'Octubre)	9	19,332	4.38%	25.71	26
62	MC Castellón (Rafalafena)	10	26,170	5.93%	34.81	35
63	MC Castellón (Casalduch)	11	25,072	5.68%	33.35	33
657	MC Castellón (Palleter)	11	21,647	4.90%	28.79	29
61	P Castellón (Pl. Constitución)	11	9,353	2.12%	12.44	12
Department 3: La Plana						
50	MC Betxí	2	6,221	1.41%	8.27	8
51	MC C.S.I. Borriana	3	31,699	7.18%	42.16	42
74	MC Nules	5	14,646	3.32%	19.48	19
78	MC Onda	6	25,520	5.78%	33.94	34
93	MC La Vall d'Uixó I	8	24,268	5.50%	32.28	32
696	MC La Vall d'Uixó II	8	9,197	2.08%	12.23	12
98	MC Villarreal I "Bóvila"	9	19,909	4.51%	26.48	26
99	MC Villarreal II "Carinyena"	9	24,691	5.59%	32.84	33
Department 4: Sagunto						
105	MC Almenara	1	5,643	1.28%	7.51	8
124	MC Seborbe	8	10,203	2.31%	13.57	14
Department 11: La Ribera						
332	MC de Benifaio	7	12,678	2.87%	16.86	17
333	MC de Almusafes	7	7,657	1.73%	10.18	10
334	Pde Sollana	7	4,907	1.11%	6.53	7
336	P de Alfarp		1,216	0.28%	1.62	2
337	P de Benimodo		1,723	0.39%	2.29	2
338	Centro de Salud de Carlet	9	15,789	3.58%	21.00	21
339	Centro de Salud de Catadau		3,227	0.73%	4.29	4
340	P de Llombai		2,112	0.48%	2.81	3
All Patients			441,526	100.00%	587.22	587

Source: the authors according to the data in the ç. * Generalitat Valenciana's population information system of the (*Regional Valencian Goverment).

Note: MC: Medical Centre; P: Practice; BH: Basic Health Area; Quota: Quota of patients assigned to a medical centre; % Quota: Percentage of patients of all the patients of the study population; N: Number of resulting questionnaires; N': Whole number of questionnaires.

The OR provides us with the measure of the association between satisfaction with the RELE system and the presence or absence of independent variables. This is the quotient between the OR with independent variables present and the OR with no independent variables:

$$OR = \frac{\left(\dfrac{1}{1+e^{-(\alpha+\beta_1 X_1 +....+\beta_n X_n)}}\right)\left(1-\dfrac{1}{1+e^{-(\alpha+\beta_1 X_1 +....+\beta_n X_n)}}\right)}{\left(\dfrac{1}{1+e^{-\alpha}}\right)\left(1-\dfrac{1}{1+e^{-\alpha}}\right)} \quad (3)$$

The OR informs us by how much the probability of a user being satisfied with the RELE system is multiplied depending on independent variables being present or absent.

When all the independent variables Xi = 0, or are absent, the OR equals:

$$OR = e^{\alpha}, \; \alpha = Ln\, OR \quad (4)$$

When independent variable Xi is present, the OR equals:

$$OR = e^{\beta_i} \quad \beta_i = Ln OR \quad (5)$$

Y quantifies the magnitude of the association between the response (patient satisfaction) and the factor of interest (age, gender, etc.) is the OR for the unit increase in variable Xi, while all the rest remain constant.

RESULTS

Our study sample comprised 35% men and 65% women, aged between 18 and 94 years. The mean age was 65. Of all the surveyed patients, 95.06% had a chronic disease (97.5% women, 94.1% men) and 67.63% had access to free pharmaceutical prescriptions.

The patients' level of education was as follows: 4.26% had no studies, 71.55% had finished primary education, 14.82% had completed secondary education, 6.47% had university studies, and 1.19% reported other studies.

Assessing Medical Prescriptions

Of all the surveyed patients, 61.6% indicated that they visited their medical centre less frequently than they did before the RELE system came into being, while only 3.7% visited it more than before. Most patients went to their medical centre on a 3-monthly basis (Figure 1). When we distinguished between chronic and acute patients, we found that the former mainly visited their medical centre quarterly (36.81%), while the latter did so occasionally (62.5%).

Most users were aware of the new RELE system, and 96.9% knew all about the treatment sheets and thought that the information on them was good (64.4%) or very good (23.9%). The majority (90.6%) were used to taking their healthcare card with them when they went to their medical centre.

When answering the question about any prescribed medication(s) being left over when they complied with treatment, 22.6% stated that some was normally left over, while 4.8% stated that they always had left over medication (Figure 2).

As for the way the RELE system operates in medical centres, 66.15% of the surveyed patients answered that their doctors always swiped their health card, 11.8% said that their doctor normally swiped it, 6.5% that their doctor did not normally do this, and only 1.5% stated that this was not the case. The RELE system has been 100% completed in these departments and only 1.2% of those surveyed receive printed prescriptions.

The time that patients spent with their doctor in their medical centre did not appear to have altered much as 71.21% of the patients thought that the time they spent was more or less the same, 15.16% said they took longer than before, and 13.63% stated that this time was shorter than before.

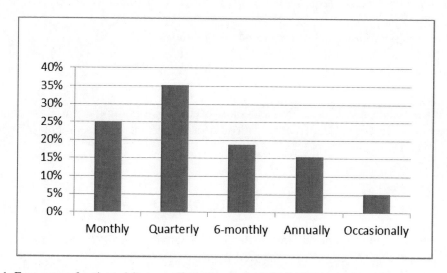

Figure 1. Frequency of patient visits to medical centres since the RELE system began.

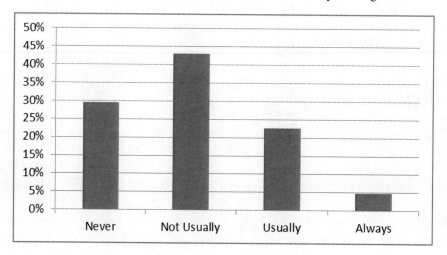

Figure 2. Percentage of responses about left over prescribed medication.

Assessing the Dispensing of Medicines

The frequency with which patients went to their chemists is shown in Figure 3. Chronic patients mostly went on a monthly basis (55.9%), and acute patients went occasionally (48.27%) or fortnightly (34.48%).

Of all the patients, 92.3% reported always taking their health card when they went to their chemists, and 72.53% always took their treatment sheets with them.

At their chemists, the health cards of 88.25% of our study population were swiped; 0.68% tended to leave their health card with the chemists and 2.41% tended to leave their treatment sheets with their prescribed medication.

Their chemists must hand them a receipt for the medications it dispenses to its customers. However, only 45.13% of our patients stated that they always received one. These receipts have to indicate the next dispensing date; 65.57% of our patients answered that this date was indicated in writing, and the person attending them also indicated it to them verbally.

When answering the question about them signing any document in their chemists' when medication was dispensed, 95.7% said they signed nothing.

To the question about taking medicines from their chemists when they knew they still had some at home, 34.41% indicated that they refused taking them, and 49.91% said they never left with more medicines. The remaining 16% accepted more medicines for the following reasons: 6.41% thought that they might not be able to acquire them when they needed them; 4.6% considered that the dispensing order would be erased from the computer system if they did not; 2.21% said that the chemists told them they had to take them even if they still had medication at home; 1.53% said they took them home because they were free.

Only 1.19% of the patients explained that their chemists told them that they were obliged to take all medications home.

Since the RELE system was set up, 82.25% of patients believed they did not take home more medication than when they previously received printed prescriptions; 15.41% said that prior to the RELE system, they did not use some printed prescriptions.

Only 50.26% of the patients knew that electronic prescriptions expired after 10 days, while the rest were not aware they had an expiration date.

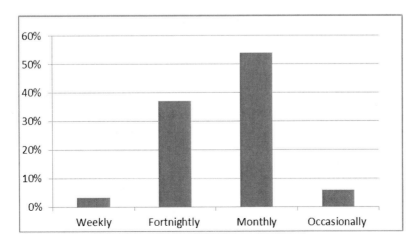

Figure 3. Frequency of visits to chemists since the RELE system started.

Assessing Use and Adhesion, and Therapeutic Compliance

Of all our surveyed patients, 94.72% took their medicines every day (men 95.6%, women 93.7%).

As for therapeutic compliance, only 56.39% said they always complied with their treatment, and 22.66% said they did so normally. However, 23.68% confessed they did not comply with it for various reasons.

Patient Satisfaction

Of all our patients, 81.9% considered that the RELE system offered them advantages as health system users, mainly because they had to visit their medical centre less frequently (72.25%), while 18% thought that the RELE system offered them no advantages. Only 11.58% were not satisfied with this system, the main reason being that they had problems with taking medication from their chemists (54.2%).

The Chaid analysis only included chronic patients (558) and showed that most users were satisfied with the RELE system. The degree of association of satisfaction with the study variables is shown in Table 2.

Table 2. The CHAID analysis results with the degree of association between each study variable and overall satisfaction

Variables	Degree of association (higher the lower p is)
Quality of information on treatment sheets	$p = 4.16504507E-19$ (*)
Fewer visits to medical centres	$p = 1.28502541E-09$ (*)
Frequency of visits to their chemists	$p = 5.41449117E-05$ (*)
Withdrawing medication even though they still had some at home	$p = 5.45556844E-05$ (*)
Withdrawing more or less medication than before the RELE system	$p = 1.53671174E-04$ (*)
Frequency medications were taken	$p = 1.60087739E-03$ (*)
Department the patient belonged to	$p = 3.81084007E-03$ (*)
Frequency of visits to medical centre	$p = 5.14203118E-02$ (ns)
Always go to the same chemists	$p = 5.76517464E-02$ (ns)
Type of system in the chemists (free or having to pay)	$p = 1.14741493E-01$ (ns)
Time with one's doctor	$p = 2.16139204E-01$ (ns)
Level of education	$p = 2.18069512E-01$ (ns)
Patient's age	$p = 3.43547762E-01$ (ns)
Time the chemists takes while dispensing	$p = 7.31662018E-01$ (ns)
Patient's gender	$p = 9.49851855E-01$ (ns)

Source: the authors. Analysed by DYANE-3©.
(*) Significant $p < 0.01$; (ns) not significant.

The profile of the most satisfied group of users with the RELE system (89 people) corresponded to the patients for whom the information on treatment sheets was good, those who visited their medical centres less frequently than before, and who did not withdraw more medication than when they previously used printed prescriptions. The most unsatisfied group comprised 22 people, of whom 68% believed that the information on treatment sheets was not clear.

By using the significant variables in the CHAID analysis (Table 2) by means of the logit model, we obtained the influence of these variables with user satisfaction with the RELE system (Table 3).

The aspects of the system that most strongly influence satisfaction with or the quality of healthcare included the information on the treatment sheets, not needing to visit one's medical centre as frequently and not spending less time with one's doctor.

Table 3. The logistic regression analysis results show the influence between the explanatory variables and user satisfaction with the RELE system

	Significance	OR ratio
Patient type (chronic or acute)	0.820	1.202
Age	0.896	1.002
Gender	0.133	1.664
Patient who finished primary studies	0.999	0.000
Patient who finished studies beyond primary studies	0.999	0.000
Patient with no studies	0.999	0.000
Type of system used in the chemists (free or having to pay)	0.715	0.849
Monthly frequency of visits to medical centre	1.000	0.000
Quarterly frequency of visits to medical centre	1.000	0.000
6-monthly frequency of visits to medical centre	1.000	0.000
Annual frequency of visits to medical centre	1.000	0.000
Occasional frequency of visits to medical centre	1.000	0.000
Fewer visits to medical centre	**0.016**	**2.413**
Quality of information on treatment sheets*	**0.005**	**3.646**
Medication left over if all medication is withdrawn	0.631	1.184
Less time with one's doctor than before the RELE system	0.118	2.381
Same time with one's doctor than before the RELE system*	**0.005**	**3.352**
More or less medication withdrawn than before the RELE system	0.420	1.382
Weekly frequency of visits to chemists	1.000	1.450E7
Fortnightly frequency of visits to chemists	1.000	4.033E7
Monthly frequency of visits to chemists	1.000	5.038E7
Occasional frequency of visits to chemists	1.000	2.177E7
Less time spent waiting in the chemists	1.000	4.992E8
Same time spent waiting in the chemists	1.000	5.158E8
More time spent waiting in the chemists	1.000	5.726E8
Taking medication daily	0.697	1.449
Taking medication weekly	0.124	0.115
Taking medication monthly	0.999	0.000

Source: the authors.
* significant variables $p < 0.05$.

Conclusion

Our results show that the RELE system operates correctly, most patients take their health card with them to their doctor and their chemists', and their health cards are swiped in most cases in both medical centres and chemist's, although legislation states that this should always be the case. A very small percentage of users leave their health card or treatment sheets at their chemists', but they should never do this. We observed that 23.68% of the patients did not comply with their prescribed treatment for several reasons.

Some studies [23] have shown that prescribing by an electronic system saves patients' time and improves healthcare. With chronic patients, however, doctors could lose control of fulfilling therapeutic treatments, so it is important to note which medications they have withdrawn from their chemists. This requirement is currently a service offered by the RELE system.

Lack of efficiency in using medications was observed as 50% of patients withdrew medication from their chemists even though they still had some medicine left over at home; this might be because they do not know some of the system's processes. We also observed that patients lacked information as only half of them knew when prescriptions expired. Moreover, 27.4% of the patients had prescribed medicines left over at home, which could result from lack of skill in the way the system is managed by doctors. This situation is dealt with by the assisted prescribing module that places alerts about dosages and durations of pharmacological treatments. Both situations could increase health expenditure, which is exactly the opposite of one of this system's objectives.

Another situation noted was that most chemists did not fulfil the obligation of handing a receipt when patients withdrew their medication as only 45% of those surveyed stated they were always given a receipt.

The aspect that most influenced patient satisfaction was the quality of the information on treatment sheets, which also influenced the fact that they visited their medical centre less, and that the time they spent with their doctor was not shorter.

In short, the RELE system, as part of the ambulatory healthcare information system in the VC, has improved the pharmaceutical prescribing and dispensing system, as reflected by the fewer patient visits to medical centres, improved accessibility and visits to chemists. Those aspects that can be improved are the content and its clearness on treatment sheets and controlling possible excess medication that patients withdraw from chemists.

The RELE system in the VC also has added value which distinguishes it from other electronic systems developed in other SACs: 1) usefulness for prescribing and dispensing stupefacient substances with a recognised digital signature [24]; 2) electronic health inspection visa; 3) prescribing from specialised healthcare centres (specialist centres and hospitals); 4) integrating chemists' local computer systems; 5) chemists can include printed prescriptions in RELE; 6) chemists can print the therapeutic plan graph; 7) prescribing non-financed products.

The current RELE system (RELE III) contains new functions like: 1) a newly designed treatment sheet, which includes collection dates, patients' diagnosis and treatment prices, and 2) replanning treatments (if a medication expires, the next lot is available) so that citizens learn to trust the system. After these changes, included to cover requirements indicated by users, re-assessing the RELE system by the Valencian Health Agency is advisable.

REFERENCES

[1] Delgado Silveira, E; Soler Vigil, M; Pérez Menéndez-Conde, C; Delgado Tellez de Cepeda, L; Bermejo Vicedo, T. Errores de prescripción tras la implantación de un sistema de prescripción electrónica asistida [Prescription errors after the implementation of an assisted electronic prescription system]. *Farmacia Hospitalaria* [Hospital pharmacy]. 2007;31:223-30.

[2] Weingart SN, Massagli M, Cyrulik A et al. Assessing the value of electronic prescribing in ambulatory care: a focus group study. *Int. J. Med. Inform.* 2009;78:571-8.

[3] Bignardi GE. Reducing prescription errors. *Lancet.* 2010;375:462.

[4] Cusack CM. Electronic health records and electronic prescribing: promise and pitfalls. *Obstet. Gynecol. Clin. North Am.* 2008;35:63-79.

[5] Alvarez Diaz AM, Delgado Silveira E, Perez Menendez-Conde C et al. Nuevas tecnologías aplicadas al proceso de dispensación de medicamentos. Análisis de errores y factores contribuyente [New technologies applied to drug dispensing process. Error analysis and contributing factors]. *Farm. Hosp.* 2010; 34:59-67.

[6] Terry M. E-prescribing: Onramp to the new electronic healthcare highway. *Telemedicine and e-health.* 2009;15:320-4.

[7] Delgado Sánchez O, Escrivá Torralva A, Vilanova Boltó M et al. Estudio comparativo de errores con prescripción electrónica versus prescripción manual [Comparative study of electronic prescribing errors versus manual prescription]. *Farm. Hos.* 2005;29:228-35.

[8] Terry K. E-prescribing. The rewards and risks. *Medical economics.* 2008; 85:26-7.

[9] Cubí Montfort R, Faixedas Brunsoms D. Viabilidad de la receta electrónica en España [Viability of the electronic prescription in Spain]. *Aten. Primaria.* 2005;36:5.

[10] Gilabert A, Cubí R. La receta electrónica en Cataluña (Rec@t): ¿Prescribimos o recetamos?. [The electronic prescription in Catalonia (Rec@t): We prescribe or prescribe"] *Aten. Primaria.* 2009;41:298-9.

[11] Ministerio de Sanidad y Política Social [Ministry of Health and Social Policy]. Las TIC en el Sistema Nacional de Salud. El programa Sanidad en Línea [ICT in the National Health System. The Healthcare Online program]. Plan de Calidad para el Sistema Nacional de Salud [Quality Plan for the National Health System]. *Estrategia 11: Sanidad en Línea* [Strategy 11: Health Online]. (consultado 25/06/2010) Disponible en: http://www.msc.es/organizacion/sns/planCalidadSNS/pdf/tic/sanidad_en_linea_WEB_final.pdf.

[12] Lerma Gaude V, Poveda Andrés J, Font Noguera I, Planells Herrero C. Sistema de alertas asociado a prescripción electrónica asistida: análisis e identificación de puntos de mejora [Warning system associated with computerized physician order: analysis and identification of areas for improvement]. *Farm. Hosp.* 2007; 31: 276-82.

[13] Gilabert-Perramon, A.; López-Calahorra, P.; Escoda-Geli, N.; y Salvadó-Trias, C. Receta electrónica en Cataluña (Rec@t): una herramienta de salud [electronic prescription in Catalonia (Rec@t): a tool health]. *Med. Clin.* (Barc.). 2010;134 Supl. 1: 49-55.

[14] Marimon-Sunol S, Rovira-Barbera M, Acedo-Anta M, Nozal-Baldajos MA, Guanyabens-Calvet J. Historia clínica compartida en Cataluña [Shared clinical history in Catalonia]. *Med. Clin.* (Barc.). 2010;134 Suppl. 1:45-8.

[15] Trillo Mata, JL. Escolano Puig, M. Villalba Garnica, P. Pascual De la Torre, M. Dominguez Carabante, A. Gestión farmacéutica en la red [pharmaceutical management network]. *Revista Valenciana de Medicina Familia* [Valencia Family Medicine magazine]. 2001;9: 39-41.

[16] Trillo-Mata, JL., Pascual-de-la-Torre, M., Perales-Marín, A., Villalba-Garnica, P., Dominguez-Carabantes, A. Sistemas de Información Sanitaria de la Conselleria de Sanitat de la Comunidad Valenciana [Health Information Systems of the Ministry of Health of Valencia]. *GAIA: Gestor de prestación farmacéutica. Farmacia de Atención Primaria* [GAIA: pharmaceutical benefits manager. Primary Care Pharmacy]. 2003;1: 14-25.

[17] Trillo-Mata, José Luís; Pedrós-Marí, Beatriz; Clérigues-Belloch, José Eduardo. Servicios de receta electrónica en la Agencia Valenciana de Salud [Electronic prescription services in the Valencia Health Agency]. Dirección General de Farmacia y Productos Sanitarios de la Agencia Valenciana de Salud [Directorate General of Pharmacy and Health Products of the Valencia Health Agency]. Sociedad Española de Informática y Salud [Spanish Society of Health Informatics]. 2008, Diciembre, n. 72.

[18] Conselleria de Sanitat [Health Ministry]. Concierto entre la Conselleria de Sanitat y los Colegios oficiales de farmacéuticos de las provincias de Alicante, Castellón y Valencia [Agreement between the Health Department and the Official Colleges of Pharmacists of the provinces of Alicante, Castellón and Valencia]. En: *Portal de la Consellería de Sanitat 2004* [Portal Department of Health in 2004]. (Consultado 28/03/2011). Disponible en: http://www.san.gva.es/docs/farmacias/1concierto.pdf.

[19] Conselleria de Sanitat [Health Ministry]. Convenio de colaboración entre la Agencia Valenciana de Salud de la Conselleria de Sanitat y los Colegios Oficiales de farmacéuticos de las provincias de Alicante, Castellón y Valencia para el desarrollo de la "atención farmacéutica electrónica-dispensación electrónica," en el marco de la receta electrónica de la Generalitat Valenciana 2007 [Cooperation agreement between the Valencian Health Agency of the Ministry of Health and the Official College of Pharmacists in the provinces of Alicante, Castellón and Valencia for the development of "e-dispensing pharmaceutical care electronic" in the framework of the electronic prescription of the Generalitat Valenciana 2007]. (consultado 28/03/2011). Disponible en: http://www.san.gva.es/docs/farmacias/2conveniodesarrollo.pdf.

[20] Conselleria de Sanitat [Health Ministry]. Convenio implantación de un modelo de dispensación en los servicios de receta electronic [Convention implementing a model of dispensing electronic prescription services]. Documento anexo para la implantación de un modelo de dispensación y facturación en el sistema integrado de receta electrónica de la Conselleria de Sanitat de la Generalitat 2008 [Annex document for the implementation of a model of dispensing and billing integrated electronic prescription of the Health Department of the Generalitat 2008 system]. (consultado 28/03/2011). Disponible en: http://www.san.gva.es/docs/farmacias/3convenioimplantacion.pdf.

[21] Ley 29/2006, de 26 de julio, de garantías y uso racional de los medicamentos y productos sanitarios [Law 29/2006 of 26 July, on guarantees and rational use of medicines and health products]. BOE núm. 178, de 27 de julio de 2006:28122-65.

[22] Santesmases Mestre M. Dyane versión 4: Diseño y análisis de encuestas en investigación social y de Mercado [Mestre Santesmases M. Dyane version 4: Design and analysis of social surveys and market research]. Madrid: Editorial Pirámide; 2009.

[23] Lapane KL, Rosen RK, Dubé C. Perceptions of e-prescribing efficiencies and inefficiencies in ambulatory care. *International Journal of Medical Informatics*. 2011; 80(1):39-46.

[24] Trillo Mata, JL; Muelas Tirado, J.; Navarro Gosalbez, M; Pérez Díaz, C; La receta informatica y/o electrónica de estupefaccientes en el ámbito de la asistencia sanitaria pública de la Comunitat Valenciana [Informatics and/or electronic recipe estupefaccientes in the field of public health care in Valencia]. *Revista de ordenación y control de productos farmacéuticos* [Journal of management and control of pharmaceutical products]. nº 2. Junio 2009.

In: Modeling Human Behavior: Individuals and Organizations ISBN: 978-1-53610-197-3
Editors: L. Jódar Sánchez, E. de la Poza Plaza et al. © 2017 Nova Science Publishers, Inc.

Chapter 9

MODELLING HUMAN BEHAVIOURS BY SHAPING ORGANIZATIONAL CULTURE

Mateusz Molasy[1]*
Faculty of Mechanical Engineering,
Wrocław University of Science and Technology,
Wrocław, Poland

ABSTRACT

At the beginning of the chapter the link between human behaviours and the culture of organization has been presented, as well as the concept of the culture of organization and its functions. The main part of the chapter discusses the methods of shaping the elements of the culture of organization – according to chosen authors as well as according to the research that has been made in four automotive enterprises in Poland. In each of the Poland's local departments of the organizations, the concerns' organizational culture was created from scratch, using more or less consciously certain rules that has been discovered and described. At the end of the chapter the set of seven measures was proposed, aimed at shaping organizational culture. They are based on the collected theoretical material and the conducted research.

Keywords: culture of organization, shaping the culture, organizational values, human resources management, human behaviours

INTRODUCTION

In times of increasing competition and increasingly strong market fluctuations measurable growth factors, such as technology, cease to work. More frequently, we look for qualitative factors closely related to man - his behaviour in the organization as well as his fit

---

* Mateusz Molasy: Wrocław University of Science and Technology, Faculty of Mechanical Engineering, 5 Łukasiewicza St., 50-371 Wrocław. Email: mateusz.molasy@pwr.edu.pl.

into it. The behaviour of people in the organization is dictated by the influence of both external and internal factors that affect every single person. Among the internal factors particularly relevant seem to be attitudes, beliefs and values of people. They justify their actions and this is where we look for employees' reactions, both favourable and unfavourable for the organization. Many authors shown the relationship between personality traits, employees' values and their organizational behaviour. Sears and Hackett (2011) as well as Cable and Judge (1997) noted that people and organizations function most effectively if their personalities, values and needs are consistent with each other.

Researchers analysing organizational behaviour often reach for the support of all human sciences. Organizational culture, combining the achievements of both psychology and sociology s well as anthropology, is undoubtedly the most important factor affecting organizational behaviour (Schermerhorn, Osborn, 2011). The importance of organizational culture is currently the subject of lively discussions and considerations on the basis of science, as well as from the point of view of management practice. Organisational culture, through its influence on the behaviour patterns of organization's members is involved in shaping other elements of the organization and management, such as strategy, structure, leadership style, and organizational learning. The configuration of individual elements of these and other areas of management emerges precisely from the way in which employees and managers perceive organizational reality of and how they behave in it.

In numerous organizations certain activities are currently being taken out, the aim of which is to increase cultural awareness. Other companies since their inception have designated cultural values and then all processes are scrupulously subordinated to them. What kind of actions to choose, however, to create the organizational culture effectively? What means and tools to reach for? How to choose and subsequently keep employees, who will be well "matched" to the organization? Getting the answers to these questions is the theme of this chapter.

ORGANISATIONAL CULTURE AND ITS FUNCTIONS

Until the mid-eighties of the twentieth century, the theme of organizational culture did not attract much of researchers' interest. It is believed that the main reasons for this state of affairs was the ordinary underestimation of the cultural factor in organizational behaviour, perception of culture as a vague element of an organization or identifying it with the professional roles or rules of the game (Bate, 1984). In 1964, Blake and Mouton (1964) used the term organizational culture to describe leadership styles. In the later period, the thinking about an organization as a culture has started, and popular literature has made of organizational culture a kind of recipe for success.

The most commonly occurring in the literature definition of organizational culture defines this concept as *"a pattern of shared basic assumptions that was learned by a group as it solved its problems of external adaptation and internal integration, that has worked well enough to be considered valid and, therefore, to be taught to new members as the correct way to perceive, think, and feel in relation to those problems"* (Schein, 2010). Lots of other popular definitions of organizational culture bring this concept to one colloquial phrase: "in our organization it is done this way" (Lundy, Cowling, 1996).

The essence of organizational culture is a set of unwritten and unspoken, but generally well-known assumptions. Certain ways of thinking are accepted and shared by a certain group. The central parts of the organizational culture are, however, the values. They reflect the person's or the organization's sets of beliefs and assumptions. They also serve as the basis of the norms that underlie behaviour (Pasher, Ronen, 2011). Herb Kelleher, founder and long-term manager of American Southwest Airlines, stated simply "Culture is what people do in the organization when no one is looking at them" (Freiberg, Freiberg, 1998). He specifies how people automatically behave, think, act and relate to each other every day: "Management leads culture and culture leads behaviour" (Dygert, Jacobs, 2004).

Organizational culture fulfils specified functions. Czerska (2003) presented them in the most synthetic way- in the form of two functions:

- Culture reduces uncertainty,
- It builds the identity of the organization.

The uncertainty stems from the variability of both the environment and internal operating conditions, and through the existing values, traditions, philosophy, history, name, brand logo and the entirety of distinctive characteristics, given organization is unique and unrepeatable. Culture for an organization is what the personality is for a man: hidden, but a unifying force, giving meaning, direction and mobilization (Kilmann, Saxton, Serpa, 1985).

Kouzes and Posner (2010) write about the functions, performed in the company through values, being an element of organizational culture, being an element of organizational culture. In their opinion, common values promote, among others, a high level of corporate loyalty and facilitate consensus on the main objectives of the organization. In addition, they induce ethical behaviour and reduce stress. Stachowicz-Stanusch (2007) notes, however, that the values are the basis of a contract optimizing human actions and building a bridge between the employee and the company. Bellou (2010) showed an important feature of organizational culture - according to the author, the shape of organizational culture has a dominant influence on the level of job satisfaction, and thus on the efficiency of everyday tasks' performance by the employees in the organization.

THE METHODS OF SHAPING ORGANIZATIONAL CULTURE

Although many organizations have ambitions to possess developed culture, it should be kept in mind that the culture cannot be simply copied.

Organizations represent independent cultural communities and each organization develops a specific culture for itself, thanks to which it gains its identity and individuality. Numerous authors of works in the field of organizational culture on the basis on fragmentary models of creating organizational culture showed that culture can be used purposefully and shaped according to the needs. Some authors focus on building culture through using personnel management or on its individual components, and others refer to other selected areas of management.

One of the best-known and most comprehensive models of creating organizational culture was presented by Liker and Hoseus (2010) on the basis of research conducted in the automotive company Toyota. According to the authors, the culture of Toyota is based on two key value pillars of the company: continuous improvement and respect for people. The model of creating culture at Toyota, according to the authors, consists of the following steps:

- Attracting competent and talented employees,
- Developing employees,
- Engaging employees in the process of continuous improvement,
- Inspiring to commitment to the company, family and society.

The process of creating the culture at Toyota is supplemented by support measures, including, among others, stimulating learning by working groups, care about clean and tidy workplace, respectively built internal communication.

Sheridan (2012) presents a fairly complex method of creating "magnetic culture" distinguishing feature of which is the maximum involvement of employees in the organization. These employees are innovative, open-minded, identify themselves with the organization's mission, its vision and values, they perform work for the organization beyond their responsibilities, optimistically foresee their future and are proud promoters of the organization in the environment. In addition, their attitude "infects" workers showing less commitment and new people in the organization. The author argues that the basis for creating magnetic culture is properly conducted recruitment, consisting of the following four stages:

- Finding the best person for the job, characterized by specific features,
- Familiarize the candidate with the culture of the organization,
- Familiarizing the candidate with the future workstation (before he is employed),
- Adaptation of the newly adopted employee,
- Ensuring the organization's new employee exactly what he has been promised in the job offer.

Besides recruitment, among the essential elements supporting the creation of magnetic culture, Sheridan mentions clear and open internal communication and special concern for the fate of the employees. It manifests itself, among others, in allowing them to do things in which they feel the strongest and to realize their own, individual paths of career development.

Aniszewska (2007) proposes the process of organizational culture's creating based on, among others, the following elements:

- Appropriately formulated mission and philosophy of the company,
- The process of employees' selection (the selection, according to the author, includes the use of techniques enabling to determine the extent to which candidate has the desired personality trait, to what extent he can identify with the organization and its objectives and the extent to which he is able to assimilate organization's standards),
- Evaluation of employees (base of which should also result from norms and values of organizational culture),
- Careers' management,
- Training and development,

- personnel marketing personnel (shaping internal company's image through the appropriate presentation of the successes of the organization, appropriate presentation of the norms and values preferred in it, which boils down to the promotion of certain attitudes preferred by the organization),
- communication process.

Rhoades (2011) believes that the creation of organizational culture should be based primarily on the construction of a particular model of employing people in the organization - only those who share the desired values and every day live the values of the organization. A useful tool might be a behaviour-based interview. It facilitates perception, already at the stage of recruitment, such behaviour in a candidate that we would like to see every day in the organization. The next step, equally important to Rhoades, is to create a system of employee assessment that to the greatest extent rewards behaviour consistent with the organizational culture. Measures should also focus on the specific concern for the fate of the employees in the organization. It is possible to use programs of organizing leisure of employees outside the company and build rich social facilities.

The subject of organizational culture creating methods is referred to by Gibbins-Klein (2011). The author describes how to create such an organization culture that fosters the leadership based on efficient achieving the goals, commitment, authority and longevity. The process of creating culture according to Gibbins-Klein starts from the top of the organizational structure, and should be headed by a person representing all the values desired in the organization. Then, in a cascade mode, the desired values should be inoculated at the lower levels of the organization, using among other things the tools of human resource management.

The need to hire employees whose beliefs and behaviour are consistent with those of the organization is also emphasized by Weyland (2011). The author shows that mutual adjustment contributes to productivity growth, and ultimately to the market success of the organization.

From these considerations it must be concluded that properly chosen methods of human resource management have a dominant influence on the shaping of organizational culture. Nevertheless, organizations should also take actions in other areas, not remaining without significance for the shaping of culture. All the time, however, it is not clear how to create the desired organizational culture comprehensively. Given the diversity of attitudes among the theorists, empirical studies were undertaken, the goal of which was to identify practical actions taken in this field within organizations.

SHAPING ORGANIZATIONAL CULTURE IN PRACTICE

To carry out the research on activities shaping organizational culture and thus approaching a holistic view of the process, inductive research strategy was used (Czakon, 2006). It is a way of testing, in which on the basis of empirical data generalization is used, and the order of the research process is based on the model of induction including, among others, the selection of cases, conducting field research, data analysis and shaping and formulating generalizations.

The conducted study took the form of field research (Kostera, 2005). In it, the researcher, during his presence in the field, all the time interprets what he observes, experiences or what he can hear from others. The collected data he records in the form of notes from the site, which facilitates systematizing the material and re-confrontation with earlier ideas (Rosen, 1991). Field studies are also particularly suitable to explore the attitudes and behaviours that can be best understood in their natural environment, in contrast to the somewhat artificial conditions of the experiment and survey (Babbie, 2012).

The study covered four companies selected from the automotive industry located in Poland. Each of them produces components and subassemblies used for the construction of vehicles, which are subsequently supplied to the market for B2B for a client or clients producing cars. The plants are local units belonging to global companies and successfully perform on the market. They passed with no problems through the automotive crisis a few years ago, year after year increasing their production and employment. In each of the Polish plants, the concerns' organizational culture was created from scratch, using more or less consciously certain rules. During the study, the everyday work, recruitment and evaluation conversations were observed and the interviews with employees at various levels were conducted; additionally, a number of organizational documents were analysed. The research, after the arrangement of the collected data, gave an image of organizational culture of individual companies and the ways of creating it.

The essential differences between enterprises chosen for the research have been presented in Table 1.

Enterprise A

Basic assumptions represent the acquisition of employees in the company as active representatives of organizational reforms. It is assumed that employees are assets of the company, not costs. All employees, regardless of position, are required to manifest an initiative, self-confidence, ability to communicate, ability to work in a team, ability to resolve conflicts, ability to improve their qualifications and learning as well as the ability to bear responsibility.

Table 1. The essential differences between enterprises chosen for the research

Enterprise	Location of headquarter	Manufactured products
A	Germany	Car's engines
B	Great Britain	Elements of suspensions
C	Germany	Electrical harnesses
D	Japan	Compressors for automotive air conditioning

Formation of organizational culture in enterprise A begins at the stage of selection of candidates for the job. Recruitment is carried out in stages, checking not only if a potential employee manifests physical abilities and has the skills necessary to take up the position, but also how he sees the values in force in the company. After going through the recruitment process shaped in this way, only those workers are employed whose own hierarchy of values fits into the system of the company. In the first week of work each employee, regardless of position, which will be entrusted to him, gets carefully familiarized with the functioning of the whole factory, from the subsequent elements of the manufacturing process to the roles of the different administrative units.

Periodic evaluation of the employee is being set not only for the fulfilment of the employee's duties, but also in terms of the extent to which he shares cultural values of the organization. The emphasis is put on the employee's awareness of the fact that he will benefit from sharing the core values - in the case of successful evaluation he may in fact take into account the promotion or financial satisfaction. Training needs resulting from the assessment are intended not only to develop the skills of employees, but also to strengthen their sense of values and increase the degree of sharing them. For the planning of training the plant uses "Training and qualifications matrix" where competences possessed and competences desired are indicated. It takes into account the degree of adjustment to the organizational culture and is used to plan possible corrections in this area.

Next steps towards shaping desired organizational culture are undertaken here by means of appropriate organization of the factory space. Even the building itself was designed from scratch and built in such a way that it not only demonstrates with its appearance and functioning the modernity and friendship towards the environment, but actually functions in a cost-effective and environmentally friendly way. There was used, inter alia, modern air ventilation system, which if desired and during changing weather conditions enables efficient recovery of heat. In order to keep low noise levels and minimize environmental pollution, a ventilation system directly at the machines was used and modern filter devices were installed. This prevents additional contamination of the premises, equipment, lighting and floors, but also minimizes the pollution emitted outside the factory. As the natural light is the healthiest for a man, modern architectural and building solutions in the plant enable optimum use of daylight. A pond surrounded by vegetation in which some fish live and an aquarium in the heart of the production hall are the examples of well-thought concept of coexistence of industry and nature.

The carrier of the concern's culture in the enterprise is also suitably shaped internal communication, which consists of, among others, the factory newspaper issued once a month. It contains (in addition to the content of automotive-related issues) articles devoted to beyond-organizational lives of the crew. Periodically, it publishes the pictures of staff, describes their personal successes, important events in their lives, hobbies, information about who got married, who had a child, etc.

The company takes actions towards the immediate environment by providing financial support for the region in which it operates - subsidizes the events of a social, cultural, sports and religious nature, but only on a local scale.

Enterprise B

Selection of candidates for jobs is in the plant based on defined "silhouette" of a desired employee. They look for workers who are, among others, solid, creative, able to work in a group and sensitive to safety in the work environment.

The basis for selection is the test of technical skills, verifying the quality of the candidate's work and sturdiness. The further course of the recruitment process consists of at least two-part interview in which it is possible to verify in detail possessing of another desirable personality traits. As an important criterion there is also considered a "good impression" that the person being recruited makes on future superiors. It is recognized that this irrational "thread of sympathy," which can occur is a good promise of the employee's adjustment to the "mood" reigning in the organization.

The company employees are assessed on the basis of defined "skills matrix," separate for each grade. An element of this assessment is a group of desirable professional behaviours that an employee should manifest. Professional behaviours stem directly from the desired values. Achieving appropriate, percentage value in this part of the assessment, allows a worker to get a rise in salaries. The company, in addition to general and specialized trainings, trains workers, among others, in the field of pro-environmental attitudes. This stems from the company document "Employment practices" which states that the specific grounds of natural resources' protection guide the company in its activity.

The idea of inducing creativity among employees is implemented in a rationalizing system. In each of the organizational units the leaders encourage to the development of innovative solutions and provide any assistance in this regard. For new technological or organizational solution the employee can be awarded with a bonus, reaching even double monthly salary.

In the company it is possible to observe particular care for employees. On the premises there is a workers' canteen. There the subsidized meals in the cashless system can be ordered for which the workers may pay using electronic workers' cards.

Enterprise C

Here, seeking for employees, they look for people with a minimum primary education, ability to work in groups, with manual skills, characterized by accuracy and availability. In addition, desirable features are initiative, good manners, communication, consistency of purpose and perceptiveness.

Candidates for the job first are first checked through written tests on the speed of response, accuracy and perceptiveness. Then, an expanded and formalized interview (which is based on specific questions to a candidate for a job to be asked) which is aimed to gain the answers regarding, among others, the overall openness of the candidate and his knowledge of the company, a real desire to work, communication skills, an initiative shown, persistence in the pursuit of goals, his hierarchy of needs, interests, non-professional interests and breadth of intellectual horizons and his attitude to life in the community. After the interview, each candidate goes to manual testing. There are five of them, and each of them corresponds to the actual stage of production line. The final stage of recruitment is six-hour training, in which the people being recruited are observed and their behaviours assessed.

The basis of the system of employee assessment is the evaluation interview, which brings the desired information on employees' satisfaction with their work. In addition, the subject of the assessment is the quality and reliability of performance, commitment and independence and the degree of adjustment to working in a group. The result of the evaluation is to determine in which areas there is a need to build career paths for individual employees.

The realization of the idea of innovation and collaborative action takes place in two independent rationalization systems. One of them is based on the possibility of submitting copyright group applications of employees' initiatives, and the second is the concern's improvement process, in which the problem to be solved is indicated from above and to its solution the working groups are set up.

The company takes care of employees and wants to be associated with them for many years. It organizes, among others, daily factory's transportation for the workers living in other municipalities than the one in which the plant is located. It attaches great importance to sports activities – the employees twice a week have an opportunity to play basketball or volleyball and for this need there is a sports hall rented. There are regular interfactorial football competitions.

Enterprise D

The selection process of new employees to the company is based on a defined silhouette of an ideal employee. Details are provided by the company's rule which says "Teamwork of employees within the organization is the basis for the prosperity of the company" and other organizational documents. They show that the company looks for employees, who are characterized by such features as: openness to people, challenges and innovation, punctuality, entrepreneurship, willingness to take risks, sustainability, communication, responsibility and willingness to continuously develop and improve their qualifications.

From the interview - the first stage of recruitment - it is concluded, among others, on candidate's activity, entrepreneurship, communication skills and priorities in personal and professional life. The next stage, checking manual skills, includes the tasks of self-assembly of a single component of the finished product assembled in the company and defining its parameters based on the use of available measurement tools. The recruiters discreetly observe the way in which the recruit performs the task. There is no time regime as the speed is regarded to be less important than accuracy and logic thinking. After completing the task the candidate is asked about self-esteem: whether the job has been done right or wrong. It is not allowed to answer "rather well," "probably well," "not so much," etc. The candidate must specify himself unequivocally, because the company, which places great emphasis on the quality of their products, cannot produce elements for which there is no assurance of the highest quality. If the employee done his job badly, but he judged himself negatively noticing his mistake, he can perform the task again.

The company defines the purpose of individual appraisal system as a "guarantee of shaping the desired behaviours of all employees, by providing them with feedback about their behaviour, attitudes and functioning of the organization." The rating allows therefore an estimate of the employee's own contribution not only to the objectives and development of the company, but also his degree of cultural adjustment.

The company takes care of the working conditions. In the production hall there are mounted so-called relaxing capsules, which are air-conditioned containers with installed air filters and sound-insulated sidewalls. The employee spending in the capsule his break at work can comfortably relax and regenerate. It is in fact quiet, and the circulation of fresh and refreshing air helps to relax.

In addition, they cherish the traditions that tie employees to the organization. An example is the annual boisterous celebration of Christmas Eve, with live music and catering. The President personally serves wine to employees and full dress is required. This event and other relevant in the life of the company are documented in the chronicle.

SUMMARY OF RESEARCH RESULTS

The result of the research showed that companies build their organizational culture mainly through actions based on human resources' management: defining the silhouette of a desired candidate in the recruitment process, deliberately planned implementation of the employee to the organization, the criteria in the periodic evaluation, selection of employee trainings, offered social package as well as other activities The individual components of these activities include not only measurable results achieved by the worker and the quality of his work, but also the fulfilment of the adopted competences, the employee's thoughts, his feelings and behaviour in relation to the results achieved, to the organization, to his own place in the structure of the company.

In the companies, instil of the organizational culture usually begins at the stage of recruitment, through which only those workers are employed whose own hierarchy of values fits into the system in force in the company and who exhibit such behaviour, which the organization expects. The first phase of recruitment is to dismiss potential employees extremely mismatched to the value system of the company. However, there are not many of such people - to the enterprise, thanks to the actions undertaken before the start of the selection, job applications are submitted primarily by the applicants who meet the criteria outlined in the job advertisement. Next selection of candidates, carried out in stages, examines not only the "hard" physical abilities of candidates, but also the "soft" ones – the way in which they match to the elements of the corporate culture. In the early days or the first week of work each employee, regardless of position, which will be entrusted to him, learns about the plant operation, the processes taking place in it and the way of working.

The periodic evaluation of employees is conducted not only from the perspective of fulfilling the obligations but also in terms of the degree to which employee shares the concerns organizational culture and the extent to which he was able to "soak" with it. In case of successful assessment of a worker, he can count on either the advancement or financial satisfaction. The emphasis is put on the employee's awareness that he can benefit from sharing the crucial values. The training needs resulting from the assessment are intended not only to develop the skills of employees, but also to strengthen their sense of value and the level of sharing them.

The staff is well cared for through offering rich social facilities and an opportunity to spend time together outside work.

In particular companies additional actions are taken, not related to the management of human resources, which also build the desired organizational culture. These activities can include properly designed physical system of the plant, nurture of occupational rituals or deliberate system of internal communication. Organizational culture can be supported also by appropriately shaped systems of rationalization. They actually to the greatest extend reward those ideas that have the most significant impact on improving or strengthening in the scope of desired attitudes and behaviours of employees.

Figure 1. The set of actions for shaping the culture of organization.

A SET OF ACTIONS AIMED AT SHAPING THE ORGANIZATIONAL CULTURE

Based on conducted empirical research, including elements of creating organizational culture, the following set of actions aimed at shaping the organizational culture was formulated:

1. Development of model organizational culture.
2. Determining the "silhouette" of a desired employee.
3. Recruitment and selection.
4. Cultural adaptation of a new employee.
5. Pro-developmental assessment of the worker.
6. Motivation through special care for employees.
7. Actions from outside the area of human resources' management.

The set of actions for shaping the organizational culture has been presented on the Figure 1.

To instil the desired organizational culture effectively in the company, first a culture pattern must be developed. Then, companies should determine the "silhouette" of a desired employee, which consists of:

- His personal characteristics (what can be included in the usual personal questionnaire and confirmed by the documents or certificates, e.g., age, place of residence, finished schools and trainings, professional experience),

- "Hard" skills (actual knowledge and ability to perform certain tasks),
- "Soft" qualities and skills (attitudes, behaviours, personality traits, revealed attitude to work, to the environment - having a decisive influence on building organizational culture).

Subsequent actions are taken in the context of recruitment and selection. It is worth considering first the use of internal recruitment - an employee, who has been staying for some time in the organization, but at a different position, could have "soaked with" organizational culture. However, if there is no possibility to employ a candidate within internal recruitment, external recruitment should be used. In this respect, the advertisement regarding recruitment to the organization should be developed, taking into account particular features: careful selection of the contents, planning its appearance and the choice of appropriate communication channels to distribute it. The advertisement itself should now attract only such candidates that we want to see in the organization.

Whether the candidate possesses the characteristics specified in the "silhouette," can be verified by analysing the submitted job application. "Hard" features can be verified in manual tests and technical exams. They are chosen and applied according to the character of work that would be performed by the new employee. The best seem to be those which possibly the best reflect the tasks performer at the future work stand or which are even the same. "Soft" qualities and skills of the candidate can be best verified in an interview. Questions asked in the middle of it better help to understand the true nature of the achievements and experience, so the companies should carefully prepare a framework for interviews - for example, a set of questions to be asked by the recruiter. Based on the answers regarding the past achievements of the candidate and the objective assessment of his earlier career it is possible to conclude about his potential future behaviours.

Another contact with the elements of the organizational culture and increasing awareness of them, the new employees have during special implementation period to the company, which takes into account full adaptation, also its cultural aspect. Implementation trainings are meant largely to integrate the new employee with the organizational culture and they should be designed regarding this purpose.

The next step in creating culture is cyclical assessment of an employee that should regularly verify matching of human resources to the predetermined silhouette of a desired employee. It should not take into account only measurable results achieved by an employee on his position and the quality of his work, but also an assessment of what the employee thinks, feels, and what actions he takes with regard to his own results, to the organization, to the environment, to the extent of his identification with the organization's culture, to his place in the structure of the company. This assessment should be pro-developmental in nature for employees and in a special way rewarding behaviours compliant with the spirit of values. It can be carried out in the form of a assessment conversation based on the self-evaluation prepared in advance by the employee.

Based on the evaluation results and difficulties identified, activities should be designed aimed at improving performance and building development path. A good and effective practice seems to be development of schedules of active and comprehensive trainings, covering also the trainings helping in shaping appropriate attitudes and behaviours.

During creating organizational culture a special concern for the fate of the employees should be considered as the primary source of motivation. It embodies, inter alia, in activities in the field of building an adequate offer of social facilities, the sense of security thanks to the working conditions and planning beyond-organizational life of workers. Thanks to that action, the employees are stronger tied to the organization and the risk of leaving work by them is reduced. These activities are accompanied by the assumption that after the use of the earlier steps of the method of creating organizational culture. We employ workers who share organizational values and who are soaked with them.

Created organizational culture to some extent is also affected by other activities from outside the area of human resources management. Work should be organized and carried out in such conditions so that the employee does not feel onerous and is able to cope with requirements posed on him without harm to his strength and health. In this regard helpful may be properly designed workplace - not only convenient and safe but also inviting, with properly chosen colours, the harmony of spatial solutions, etc. Work carried out in conditions of comfort is not only more efficient, but at the same time the body does not take unnecessary losses related to adaptation to difficult environmental conditions.

Not without significance is the rationalisation systems used that enhance the creativity of employees and their willingness to make special efforts and take voluntary additional measures for the organization.

CONCLUSION

Based on the collected theoretical material and the conducted research, a set of seven measures was proposed, aimed at shaping organizational culture, one of the most important management areas affecting the behaviour of the organization. Most of the actions are closely related to human resources management.

The possible main advantages of shaping the organizational culture according to the actions proposed should include the success of the organization on the labour market - attracting the best candidates and being able to prevent them effectively from giving up their work. The surveyed companies do not exhibit worrying phenomena of excessive fluctuation of personnel or problems with finding employees to new positions. Among the fundamental flaws and weaknesses, however, can be listed required high qualifications of personnel department employees in terms of performer work and excellent interpersonal skills of middle and senior management. The method is also characterized by long implementation process.

Undoubtedly, the evidence of success of their activities they find in the automotive industry - thanks to the results achieved by the company that build their culture in this way. As far as they prove to be effective in other industries and non-productive organizations, may show potential further research in the future.

REFERENCES

Aniszewska G. (2007), Kultura organizacyjna w zarządzaniu [Organizational culture management].
Babbie E.R. (2012), The Practice of Social Research.

Bate P. (1984), The Impact of Organizational Culture on Approaches to Organizational Problem-Solving, *Organization Studies* 5/1, 43–66.

Bellou V. (2010), Organizational culture as apredictor of job satisfaction: the role of gender and age, *Career Development International* 15 No. 1, 4–19.

Blake R.R., Mouton J.S. (1964), The Managerial Grid.

Cable D.M., Judge T.A. (1997), Interviewers' Perceptions of Person-Organization Fit and Organizational Selection Decisions, *Journal of Applied Psychology* 82 No. 4, 546–561.

Czakon W. (2006), Łabędzie Poppera – case studies w badaniach nauk o zarządzaniu [Swans Popper - case studies in management sciences research], *Przegląd Organizacji* 9/2006, 9–12.

Czerska M. (2003), Zmiana kulturowa w organizacji [Cultural change in the organization]. Wyzwanie dla współczesnego menedżera [The challenge for the modern manager].

Dygert C.B., Jacobs R.A. (2004), Creating a Culture of Success: Fine-Tuning the Heart and Soul of Your Organization.

Gibbins-Klein M. (2011), Winning by thinking: how to create a culture of thought leadership in your organization, *Development And Learning In Organizations* 25, 8–10.

Kilmann R.H., Saxton M.J., Serpa R. (1985), Gaining Control of the Corporate Culture.

Kostera M. (2005), Kultura oraganizacji. Badania etnograficzne polskich firm [Ethnographic studies of Polish companies].

Kouzes J.M., Posner B.Z. (2012), The Leadership Challenge: How to Make Extraordinary Things Happen in Organizations.

Liker J.K., Hoseus M. (2008), Toyota Culture: The Heart and Soul of the Toyota Way.

Lundy O., Cowling A. (1996), Strategic Human Resource Management.

Pasher E., Ronen T. (2011), The Complete Guide To Knowledge Management.

Rhoades A. (2011), Built on Values: Creating an Enviable Culture that Outperforms the Competition.

Rosen M. (1991), Coming to terms with the field: Understanding and doing organizational ethnography, *Journal of Management Studies* 28, 2.

Schein E.H. (2010), *Organizational Culture and Leadership*, Fourth Edition.

Schermerhorn J.R, Osborn R.N., Uhl-Bien M., Hunt J.G. (2011), Organizational Behavior.

Sears G.J., Hackett R.D. (2011), The Influence Of Role Definition And Affect In LMX: A Process Perspective On The Personality – LMX Relationship, *Journal Of Occupational And Organizational Psychology* 84, 544-564.

Sheridan K. (2012), Building A Magnetic Culture. How To Attract And Retain Top Talent to Create an Engaged, Productive Workforce.

Stachowicz-Stanusch A. (2007), Potęga wartości [the power of]. Jak zbudować nieśmiertelną firmę [How to build an immortal company].

Weyland A. (2011), How to attract people who are in sync with your culture, *Human Resource Management International Digest* 19 No. 4. 29-31.

In: Modeling Human Behavior: Individuals and Organizations ISBN: 978-1-53610-197-3
Editors: L. Jódar Sánchez, E. de la Poza Plaza et al. © 2017 Nova Science Publishers, Inc.

Chapter 10

ROBBERY ATTRACTIVENESS AMONG URBAN AREAS: A COMPUTATIONAL MODELLING APPROACH

R. Cervelló-Royo[1,*], *E. Conca-Casanova*[2,†], *J.-C. Cortés*[2,‡]
and Rafael-J. Villanueva[2,§]

[1]Department of Economics and Social Sciences,
Universitat Politècnica de València, Spain
[2]Instituto Universitario de Matemática Multidisciplinar,
Universitat Politècnica de València, Spain

Abstract

Due to the current economic situation and the growth of inequality in cities, there are several urban areas in which crime rates have increased in a dramatic way; thus, some of them have turned to be considered centers of attraction for burglars. From both the economic and social point of view it would be interesting to try to give an explanation to the problem. In order to study the robbery among urban areas, two conceptions are studied using a mathematical modelling approach: a) The probability that a robbery might take place in an urban area when considering the attractiveness (that will be measured as the number of businesses, shops, stores, etc.) in that area; b) Simulation to check how burglars will move from one urban area to another. Results depict quite well the behavior of burglars for a city like Valencia and its different urban areas and/or neighbourhoods.

PACS: 05.45-a, 52.35.Mw, 96.50.Fm

Keywords: robbery modelling, computational model, simulations

AMS Subject Classification: 60H30, 34F05, 62P20, 91,B25, 91B70

---

[*]E-mail address: rocerro@esp.upv.es
[†]E-mail address: enconca91@gmail.com
[‡]E-mail address: jccortes@imm.upv.es
[§]E-mail address: rjvillan@imm.upv.es

1. Introduction

Accordingly to the findings by Glaeser et al. [4], inequality among neighborhoods is important to explain crime in cities. In fact, crime rates are higher in unequal cities. Daly et al. [2] and Fajnzylber et al. [3] show how connection between crime and inequality is as strong within urban areas as it is across countries. Due to the current economic situation and the growing inequality in cities, there are several urban areas in which crime rates have increased in a dramatic way; thus, some of them have turned to be considered centers of attraction for burglars.

Bernasco et al. [1] assessed the effects of attractiveness, opportunity and accessibility to burglars on the residential burglary rates of urban neighborhoods. Their results suggest that all three factors, attractiveness, opportunity and accessibility to burglars, pull burglars to their target neighborhoods.

In last years, there have been a series of reinforcement strategies in several neighborhoods and districts, with a strong component of prevention and citizen safety. That is the reason of the emphasis put on the study of criminality and safety reinforcement, with the aim of increasing the residential and commercial appeal of these areas.

From both, the economic and social point of view, it would be interesting to try to give an explanation to the problem.

Some mathematical models studying the evolution of the urban crime have been proposed. For instance, Jones et al. [6], Rodriguez et al. [7] and Short et al. [8, 9] where the authors model the problem using partial derivatives equations to include the spatial factor. Then, they perform a dynamical analysis to find the equilibrium points and study their stability.

In order to study the robbery appeal among urban areas, two conceptions are studied by our mathematical modelling approach: a) the probability that a robbery might take place in an urban area when considering the attractiveness (that will be measured as the number of businesses, shops, stores, etc.) in that area; b) Simulation to check how burglars will move from one urban area to another.

With this aim, we will take as a reference the city of Valencia and its different urban areas. Valencia city is divided into 87 neighborhoods as it is shown in Figure 1.

2. Method

As it has been previously pointed out, we want to study the behavior of crime, more precisely, of burglars in the neighborhoods of the city of Valencia.

2.1. Graph and Attractiveness Models

For this purpose, we model all the neighborhoods of Valencia as a graph, in which each vertex/node represents each one of the Valencian neighborhoods and the edges/lines connect nearby neighborhoods with their surrounding ones (see Figure 2).

As Luttmer [5] states, envy and crime are more likely to be directed toward near neighbourhoods than geographically dispersed compatriots. That is the reason we have decided

Table 1. Number of surrounding neighbourhoods of the 15 Valencian neighbourhoods which limit with a greater number of nearby neighbourhoods

Malilla has 11	Penya-roja has 8
Mestalla has 11	La Raiosa has 8
Cami Real has 10	Patraix has 8
Russafa has 10	Nou Moles has 8
Benimaclet has 9	Arrancapins has 8
La Carrasca has 9	El Botánic has 8
La Punta has 9	Sant Francesc has 8
Benicalap has 8	

surface. Therefore, we have accounted for the total number of businesses (shops, stores, etc.) located in a neighbourhood divided by its surface (km^2). Thus, the fifteen most attractive neighbourhoods in Valencia are the ones listed in Table 2.

Table 2. The 15 most attractive Valencian neighbourhoods (measured by number of businesses per square kilometer). Source: Authors' own elaboration

Sant Francesc has 68.31 $buss \times km^2$	La seu has 22.48 $buss \times km^2$
El mercat has 62.54 $buss \times km^2$	La Petxina has 22.03 $buss \times km^2$
Pla del remei has 50.82 $buss \times km^2$	Arrancapins has 21.35 $buss \times km^2$
La Roqueta has 42.03 $buss \times km^2$	Jaume Roig has 21.30 $buss \times km^2$
Gran via has 32.38 $buss \times km^2$	La Bega Baixa has 21.18 $buss \times km^2$
La xerea has 30.22 $buss \times km^2$	Patraix has 20.82 $buss \times km^2$
Russafa has 27.52 $buss \times km^2$	Albors has 20.78 $buss \times km^2$
El Calvari has 23.40 $buss \times km^2$	

As we can check, Tables 1 and 2 share the following neighbourhoods: Sant Francesc, Russafa, Arrancapins and Patraix. This makes sense, since the first two neighbourhoods are included in the Central Business District Area (CBD) (whose residents have a high standard of living) of the city and the second ones as well as former conurbations and former urban growth boundaries of the city.

First of all, we should define which is the probability of a robbery to take place in a neighbourhood. In our case it will be related with the neighbourhood business density. For this purpose, a logistic function has been considered

$$P(x) = \frac{e^{ax+b}}{1 + e^{ax+b}}, \qquad (1)$$

where x is the attractiveness while a and b are two constants to be determined.

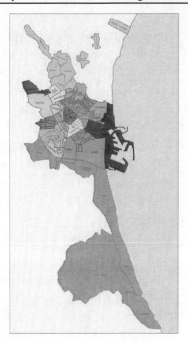

Figure 1. Map of Valencia divided into its 87 neighborhoods. Source: Valencia City Council Statistics Office.

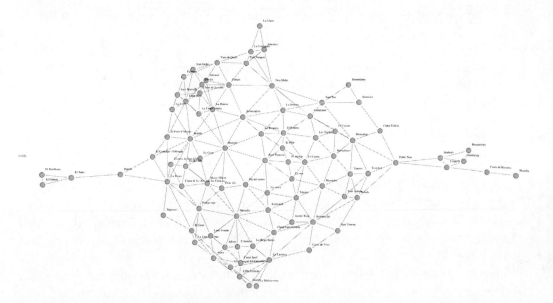

Figure 2. Graph of Valencia. The nodes represent the neighborhoods. Each edge connects two adjacent neighborhoods.

to consider the 15 more *connected* neighbourhoods. In the Table 1 we can see the fifteen neighbourhoods which limit with a greater number of nearby/surrounding neighbourhoods.

As a measure of the attractiveness, we have considered the density of businesses per

2.2. Simulation

The simulation we have carried out consists of the following procedure:

1. We choose a random neighbourhood and suppose that the burglar will start to move from this first point in order to rob in the rest of the city.

2. We will allow the burglar to do up to 12 neighbourhood changes in a simulation. This will reflect the daily time spent by a burglar trying to rob and the time periods which takes to change from one neighbourhood to another.

3. When the burglar is in a neighbourhood, we determine the probability of robbing in that neighbourhood by using the neighbourhood attractiveness x and the probability formula given by (1), $P(x)$. Then, we generate a random number $r \in [0, 1]$. If $r < P(x)$, it means that the burglar has robbed in that neighbourhood and the simulation will finish by saving the neighbourdhood where the robbery took place. If $r > P(x)$, there is no robbery and the burglar will move to the most attractive nearby/surrounding neighbourhood. If after 12 movements (neighbourhood changes) the burglar has not robbed anything, we will finish the simulation and then, start with a new one.

Our goal is to perform a lot of simulations in order to check in which neighbourhood there are more robberies.

It is also important to notice that the results of the simulation depend on the parameters a and b which appear in the probability formula $P(x)$ given by expression (1).

3. Results

3.1. Choice of a and b: Latin Hypercube Sampling (LHS)

Since the values that a and b are unknown, we have simulated several of them in an organized way. Considering the function $P(x)$, we have carried out several tests; then, we have decided that a will take values included in the range $[-10, -5]$ and b will take values lying in the range $[0, 0.1]$. Thus, considering these values, the probability $P(x)$ have taken values which were neither too high nor too low.

In order to choose 10,000 pairs of values we have considered the technique called Latin Hypercube Sampling (LHS) [10]. This technique has been applied to select sets of the variable parameters (a and b) to be substituted into the model $P(x)$ in order to perform a simulation. LHS, a type of stratified Monte Carlo sampling, is an efficient method for achieving equitable samples of all input parameters simultaneously. In our problem, by LHS we have obtained an equitable sample of 10,000 input parameters simultaneously. Then, we have substituted each set of the 10,000 parameters into the model and afterwards we have performed the simulations. The set of 10,000 results from the obtained simulations represents a wide range of feasible behaviors that we will analyze.

3.2. Simulation Details

For each a and b provided by LHS technique, we have carried out a simulation process which consists of carrying out 100 simulations for each neighbourhood, for each one of the 87 neighbourhoods, that is, a total of 8,700 simulations for each a and b.

Finally, we have saved the neighbourhoods in which a robbery took place in each one of the simulations.

3.3. Results of the Simulations

After conducting an accurate analysis of the results, only 8 pairs out of the 10,000 pairs (a, b) obtained by LHS provided robberies and they were:

$$(-9.94424, 0.0409654), (-8.50674, 0.0144289), (-7.93203, 0.034762),$$
$$(-7.54837, 0.0810752), (-7.31702, 0.0517513), (-7.28269, 0.0319738), \quad (2)$$
$$(-6.73781, 0.0374658), (-5.06928, 0.0780498).$$

In Figure 3 we can see where the pair of parameters (a, b) are located in the plane.

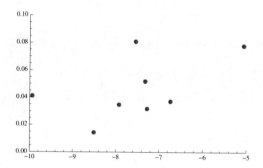

Figure 3. Graphical location of the pairs of parameters (a, b) where robberies arose in the simulations.

The total number of robberies which corresponds to each combination of a and b related above were

$$39, 30, 194, 2863, 806, 268, 687, 5714.$$

Thus, it can be observed that the number of robberies is related with a second component b high, and at the same height, with a first component a greater. As we have previously underlined, each test for each pair a and b comprises 8,700 simulations, therefore, from the values indicated above, the ones until 8,700 are simulations without robbery.

In Figure 4, we show graphically the number of robberies by neighbourhood for each one of the pairs a and b appearing in (2). Neighbourhoods are shown in X-axis and number of robberies are shown in Y-axis. In some graphs, blue points do not appear because they are values much greater than the range shown in the graph.

Below, we show only the first 5 neighbours in which more robberies have taken place and its percentage with respect to the total.

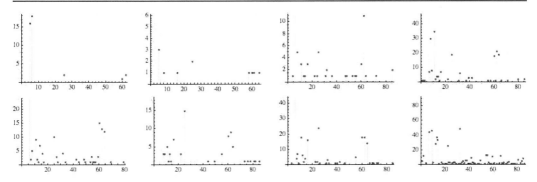

Figure 4. Number of robberies per neighbourhood. Neighbourhoods are encoded by number. With independence of the values a and b, the neighbourhoods with robberies are more or less the same in all the graphs.

- $(a, b) = (-9.94424, 0.0409654)$

    1. There have been 18 robberies in Sant Francesc, 46.15%.
    2. There have been 16 robberies in El mercat, 41.02%.
    3. There have been 2 robberies in Jaume Roig, 5.12%.
    4. There have been 2 robberies in La Bega Baixa, 5.12%.
    5. There have been 1 robbery in Ciutat Jardi, 2.56%.

- $(a, b) = (-8.50674, 0.0144289)$

    1. There have been 19 robberies in Sant Francesc, 63.33%.
    2. There have been 3 robberies in El mercat, 10%.
    3. There have been 2 robberies in Jaume Roig, 6.66%.
    4. There have been 1 robbery in Benimaclet, 3.33%.
    5. There have been 1 robbery in Ciutat Jardi, 3.33%.

- $(a, b) = (-7.93203, 0.034762)$

    1. There have been 84 robberies in Sant Francesc, 43.29%.
    2. There have been 62 robberies in El mercat, 31.95%.
    3. There have been 11 robberies in La Bega Baixa, 5.67%.
    4. There have been 5 robberies in Pla del remei, 2.57%.
    5. There have been 5 robberies in Jaume Roig, 2.57%.

- $(a, b) = (-7.54837, 0.0810752)$

    1. There have been 1809 robberies in Sant Francesc, 63.18%.
    2. There have been 851 robberies in El mercat, 29.72%.
    3. There have been 35 robberies in La Roqueta, 1.22%.

4. There have been 30 robberies in Pla del remei, 1.04%.
   5. There have been 21 robberies in La Bega Baixa, 0.73%.

- $(a, b) = (-7.31702, 0.0517513)$

   1. There have been 405 robberies in Sant Francesc, 50.24%.
   2. There have been 293 robberies in El mercat, 36.35%.
   3. There have been 15 robberies in Ciutat Jardi, 1.86%.
   4. There have been 13 robberies in La Bega Baixa, 1.61%.
   5. There have been 12 robberies in Benimaclet, 1.48%.

- $(a, b) = (-7.28269, 0.0319738)$

   1. There have been 112 robberies in Sant Francesc, 41.79%.
   2. There have been 82 robberies in El mercat, 30.59%.
   3. There have been 15 robberies in Jaume Roig, 5.59%.
   4. There have been 9 robberies in La Bega Baixa, 3.35%.
   5. There have been 8 robberies in Ciutat Jardi, 2.98%.

- $(a, b) = (-6.73781, 0.0374658)$

   1. There have been 300 robberies in Sant Francesc, 43.66%.
   2. There have been 228 robberies in El mercat, 33.18%.
   3. There have been 24 robberies in Jaume Roig, 3.49%.
   4. There have been 18 robberies in Ciutat Jardi, 2.62%.
   5. There have been 18 robberies in La Roqueta, 2.62%.

- $(a, b) = (-5.06928, 0.0780498)$

   1. There have been 2966 robberies in Sant Francesc, 51.90%.
   2. There have been 1049 robberies in El mercat, 18.35%.
   3. There have been 269 robberies in La Roqueta, 4.70%.
   4. There have been 201 robberies in Pla del remei, 3.51%.
   5. There have been 179 robberies in La Bega Baixa, 3.13%.

As we can check, there have been robberies in four neighbourhoods, *El mercat, Sant Francesc, Jaume Roig, Ciutat Jardi*, for all the pairs a and b.

Then, considering all the neighbourhoods where robberies have been occurred for any pair of parameters (a, b) in (2), we can state the following:

1. From the previous neighbourhoods, only Benimaclet, Arrancapins and Sant Francesc belong to the group of the 15 Valencian neighbourhoods that limit with a greater number of nearby neighbourhoods.

2. From the previous neighbourhoods, only Campanar, Ciutat Jardi and Benimaclet do not belong to the 15 most attractive Valencian neighbourhoods (measured by number of businesses per square kilometer).

3. The previous neighbourhoods determine four Valencian urban areas in which more robberies seem to take place:

   - The Area 1 which comprises the neighbourhoods: El mercat, Sant Francesc, Pla del remei, La Roqueta and Arrancapins, marked on the map (in Figure 5 in red) with the locations 1.5, 1.6, 2.2, 3.2 and 3.4.
   - The Area 2 which comprises the neighbourhoods: Campanar and El Calvari, marked on the map (in Figure 5 in blue) with the locations 4.1 and 4.3.
   - The Area 3 which comprises the neighbourhoods Jaume Roig and Benimaclet marked on the map (in Figure 5 in green) with the locations 6.3 and 14.1.
   - The Area 4 which comprises the neighbourhoods Albors, Ciutat Jardi and La Bega Baixa, marked on the map (in Figure 5 in yellow) with the locations 12.2, 13.2 and 13.4.

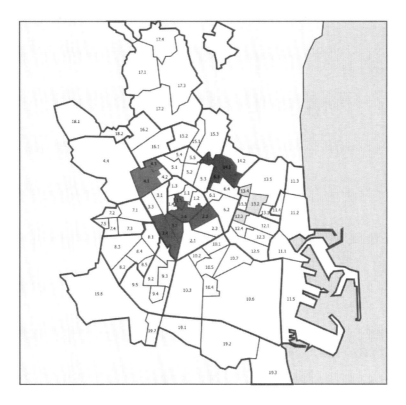

Figure 5. Map of Valencia divided into neighbourhoods. In colours, the Valencian urban areas in which more robberies seem to take place: red for the Area 1; blue for the Area 2; green for the Area 3; yellow for the Area 4.

Conclusion

In this work we have presented a computational mathematical modelling approach for the case of robbery attractiveness among urban areas. More precisely, an application of a mathematical modelling to get the probability that a robbery might take place in an urban area when considering the number of businesses (shops, stores, etc.) located in that area and a simulation to check how burglars will move from one urban area to another.

We have shown how by means of a logistic function we have been able to find which is the probability of a robbery take place in a neighbourhood, in our case it will be related to the neighbourhood businesses density. Moreover, by LHS we have been able to carry out a simulation process which consisted of carrying out 100 simulations for each neigbourhood, that is to say, a total number of $8,700$ simulations for each a and b.

For more than $10,000$ tests generated by the LHS, only eight pairs of (a, b) provided robberies; however, these values were not extremely concentrated in the same spatial place. This may imply that the model is very sensitive to the values a and b which define the probability. Nevertheless, the values of a and b have some effect on the number of robberies (quantitative) but not on the detection of the areas (qualitative).

As we have seen, there were four neighbourhoods which have been robbed for all the pairs a and b in which robberies took place. It allowed us to determine four Valencian urban areas in which more robberies took place.

To finish, we would like to emphasize that the model simulated provided us with actual results, however and for future studies, the consideration of experts opinion and different scenarios might help to improve the model and its simulations. Additional factors, like the consideration of the existing connections with the surrounding/nearby neighbourhoods apart from the density and probability might be considered in future studies. In our opinion, the case study constitutes a promising area of research in social sciences.

References

[1] Bernasco, W. and Luykx, F. (2003). Effects of attractiveness, opportunity and accessibility to burglars on residential burglary rates of urban neighborhoods, *Criminology*, 41(3), 981–1002.

[2] Daly, M., M. Wilson, and S. Vasdev. 2001. Income Inequality and Homicide Rates in Canada and the United States, *Canadian Journal of Criminology*, 43(2), 219–236.

[3] Fajnzylber, P., D. Lederman, and N. Loayza, 2002. Inequality and Violent Crime, *Journal of Law and Economics*, 45(1), 1–40.

[4] Glaeser, E. L., Resseger, M. and Tobio, K. (2009). Inequality in cities. *Journal of Regional Science*, 49(4), 617–646.

[5] Luttmer, E.F.P. 2005. Neighbors as Negatives: Relative Earnings and Well-Being, *Quarterly Journal of Economics*, 120(3), 963–1002.

[6] P.A. Jones, P.J. Brantingham, and L.R. Chayes, Statistical models of criminal behavior: The effects of law enforcement actions, *Math. Models Methods Appl. Sci.*, Vol. 20, Suppl. (2010) 1397–1423, DOI: 10.1142/S0218202510004647.

[7] N. Rodriguez and A. Bertozzi, Local existence and uniqueness of solutions to a PDE model of criminal behavior, *Math. Models Methods Appl. Sci.*, special issue on *Mathematics and Complexity in Human and Life Sciences*, 20 (2010), 1425–1457.

[8] M. B. Short, M. R. D'Orsogna, V. B. Pasour, G. E. Tita, P. J. Brantingham, A. L. Bertozzi, and L. B. Chayes, A statistical model of criminal behavior, *Math. Models Methods Appl. Sci.*, 18(suppl.):1249–1267, 2008.

[9] M.B. Short, A.L. Bertozzi, and P.J. Brantingham, Nonlinear patterns in urban crime: hotspots, bifurcations, and suppression, *SIAM J. Applied Dynamical Systems*, Vol. 9, No. 2, pp. 462–483, 2010.

[10] A. Hoare, D.G. Regan, D.P. Wilson (2008): Sampling and sensitivity analyses tools (SaSAT) for computational modelling, *Theoretical Biology and Medical Modelling* 5, article 4, 2008. doi:10.1186/1742-4682-5-4.

In: Modeling Human Behavior: Individuals and Organizations ISBN: 978-1-53610-197-3
Editors: L. Jódar Sánchez, E. de la Poza Plaza et al. © 2017 Nova Science Publishers, Inc.

Chapter 11

THE PEAK WORK OF THE PATRIARCH RIBERA IN THE COUNTER-REFORMATION: THE ROYAL SEMINARY-SCHOOL OF CORPUS CHRISTI OF VALENCIA (SPAIN)

Carlos Lerma[1],*, Ángeles Mas[1], Enrique Gil[1] and Jose Vercher[1]
[1]Universitat Politècnica de València, Valencia, Spain

ABSTRACT

This research chapter shows the figure of Mr Juan de Ribera. He was the Patriarch and promoter of one of the most important buildings of the Valencian Renaissance. We focus on his personal commitment and their behavior when built the Royal Seminary-School of Corpus Christi of Valencia.

This building is the culmination of his personal work. The Patriarch Ribera was a very influential person in the city of Valencia, who had relations with the Court of King Philip III of Spain. Ribera developed and promoted a policy of building churches and all types of religious buildings in the city of Valencia, as a conventual city. However, he ended all his knowledge at the Seminary-School of Corpus Christi. Counter-Reformation Instructions published by Saint Charles Borromeo after the Council of Trent (1545-1563) greatly influenced the Patriarch. He sacrificed his personal fortune and sought funding nobles and royalty to build this building. 400 years after of the construction, the great work of Patriarch still stands, still used for their original duties and his memory is still alive.

Keywords: Counter-Reformation, architecture, Valencia, construction, personal behavior

---

* E-mail: clerma@csa.upv.es.

INTRODUCTION

The Royal Seminary-School of Corpus Christi of Valencia is an institution focused on the formation of priests in the 17th century. The institution maintains extensive documentation on its file, as legal documents, notarial documents, incunabula and a large number of books. Its founder was the Blessed Juan de Ribera, who after the Council of Trent decide to build his own institution following the provisions of the Council. Architecture cannot be understood as an isolated event, hereby the importance of contextualizing the architecture in the historical period it was built. We treated here an historical issue, social or political issues also.

Surely, the Seminary-School of Corpus Christi (Figure 1 and Figure 2), popularly called School of the Patriarch, needs no introduction as it is one of the most emblematic building of the city of Valencia, known and respected by its inhabitants in all times. Suffice it is to say that its founder, Patriarch Juan de Ribera (1532-1611) spent his fortune in the construction and maintenance of the Seminary-School, with the aim of training priests, a task which continues today taking place.

The construction of this Seminary-School is part of the historical period of the Renaissance, in which architecture played a leading role. All provisions governing and influenced the bishops, architects and builders gathered in Instructiones Fabricae of Saint Charles Borromeo.

The Renaissance Project

Valencia was fully developed in an architectural Renaissance and was in the line with the rest of Spain although with some differences (Llopis 2002a).

Figure 1. The Seminary-School of Corpus Christi of Valencia (background).

Figure 2. Inside the church of the Seminary-School of Corpus Christi.

The Renaissance concept of the architect as an enlightened man, formed in many subjects linked more or less directly to their profession, that is, a very humanist architect of Renaissance culture, is a figure directly derived from the concept that it is clear from Vitruvian text (Llopis 1997). The process concludes with the figure of Juan de Herrera and architectural professional structure of the court of Philip II, although in Valencia the most representative figure should be Gaspar Gregori, among other buildings, worked at the Seminary-School of Corpus Christi (Llopis 2002b).

METHODS

The literature on the Seminary-School of the Patriarch, its founder or the influence of both on the back architecture of the city of Valencia is very extensive.

The amount of volumes and delay in time imply that has already approached the building from many aspects and points of view. However, only a few studies have been performed by architects who deal only this building or part of it.

In the analysis of the constructive aspects of the building it is important to highlight the historical period of its conception and building, since architecture is a reflection of the culture of each era (society, economy, construction...). The literature review is not intended to make a state of the art (Llopis 1997), but it is a comprehensive review of information relevant to this chapter, analyzing the published works have been dedicated fully or partially to the building or to the figure of its founder. We have studied the historical context of the sixteenth and seventeenth centuries in which takes place the construction of the Seminary-School. The intention is to know the reasons that prompted the Patriarch Ribera to undertake this project.

The historical period in which the construction of the Seminary-School is part influences especially, since the architecture is not without political decisions, religious... of his time.

To understand the location of the building, including its size, position of some architectural elements (like the dome, bell tower, etc.) is interesting to know the urban environment of the building and its evolution.

We also study in depth the different maps and historical maps where the Seminary-School appears. There are many, but highlight the map of Mancelli (1608), it shows us for the first time an image of the building, and Tosca (1704) to be very precise in details.

The most prominent aerial images are certainly Alfred Guedson (1858), but are general views of the city and do not represent only the building in question.

RESULTS

The construction of the Seminary-School of Corpus Christi is a part of the historical period of the Valencian Renaissance (15th-17th centuries). To understand how and why the decisions that led us to construction the building is necessary that we need to know the historical, political, social, religious, etc. context of that time.

Europe sought a new language. The universalist spirit that characterizes Europe in the 13th century is disintegrating in particular fragments in the 14th century to disappear in the 15th. The ideals that gave the Church in the first Gothic are replaced by the attitude of critical thinking developed in the recent universities, where classical Greek or Latin are read, postulates theological are discussed and people doubt dogmas. The printing press allowed the dissemination of written culture. In the Gothic building dimensions possess humanity, but in the Renaissance will be the man who dominates everything.

The Spanish society of 1500 had a very weak commercial component in most of the mainland. By contrast the nobility was holding in his hands a huge economic power, but politically they will cut its prerogatives by the absolutist state. All this explains why the Spanish Renaissance are confined to promote the dictates by the Court, the Church and the nobility.

In our case especially it higlight the continuity Middle Ages-Renaissance by the persistence of forms of power, ownership and medieval mentality in the 15th and 16th centuries.

The Spanish economic situation would have required a great austerity, that neither Philip III and his favorite the Duke of Lerma were able to take. They spent the 10% of the revenues of the tax in the royal wedding in 1599 (VVAA 1999).

The location of the Seminary-School of the Patriarch is linked, among other issues, to the University, whose situation seems to respond to a more or less explicit desire to alienate students from power centers and meeting ones. In any case far from the Market Square, a regular meeting place (Wedding 2001).

In the Renaissance multitude of religious buildings are constructed inside and outside the walled city of Valencia. A primitive parishes and mendicant orders we add now monastic buildings (1536 S. Sebastian, S. Fulgencio and Corona 1563 S. Joachim and Sta. Ana 1564 S. Juan de Ribera 1587, la Sangre 1596, Pie de la Cruz 1597, S. Gregorio 1600, Sta. Monica 1603...) altering the urban morphology in favor of a monastic city (Llopis et al. 2004). The

Gothic gives the way to Renaissance with representative examples: the closing of the Torre de la Generalitat (1516-1600), the lodge added to Seo, the Palace of Ambassador Vich (1521), the Convent of San Miguel de los Reyes (1546-1835) or the Seminary-School of Corpus Christi (1586-1615).

San Juan de Ribera was named archbishop of Valencia in 1569 and, as we know, he will promote many religious buildings in his congregation. He organizes in Burjassot construction ovens and means for making small tiles (13.5 x 13.5cm^2, size underused in Valencia because the usual was 15 x 15cm^2) and polychrome imitating the Seville ones. The baseboard of the cloister of the Seminary-School of Corpus Christi are still made with the technique of the artist that was done in Seville before the arrival of the Italians. Artist worked in Burjassot, both Sevillian artists and from Talavera de la Reina (Vizcaino 1999).

Saint Charles Borromeo (1538-1584) was cardinal and archbishop of Milan. We can highlight his figure and its role in the Council of Trent (1545-1563) in which, as a Secretary of State, leads the prior negotiation and the correspondence between Rome and Trento. When tensions between the two cities relaxed he focus its efforts on the completion of the Council. In this meeting decrees the bula of 1564 which contained his signature. In addition, as archbishop of Milan he wanted to implement, as soon as possible, in his diocese Tridentine reforms.

Borromeo had an extensive knowledge of the issues discussed at the Council and decided to publish, fourteen years after its completion, a summary of Catholic traditions regarding the design of churches. Officially the *Instructiones* were for Milan, but his intention was to have a more widespread use (Gallegos 2004).

In the present case, the Patriarch Ribera bought all the property belonging to a whole block (Lerma 2012), bounded by four streets that would keep the proper distance to other buildings. As the *Instructiones* explained it is easier to get this in cities, due to its own path, rather than in rural areas. As is known, the city of Valencia is essentially flat, but the Seminary-School was located in an area of slight slope, which would benefit when circumvent storm water runoff and overflows of the called Guadalaviar River (now Turia River). Where appropriate, the biggest problem is derived by successive floods of the river that would condition its construction, from the situation of the access to the choice of the materials of the facade. The forecast of Blessed Juan de Ribera did build the building higher than the surrounding streets; thus, the neighboring University of Valencia was badly damaged and the Seminary-School just a few centimeters of water in the big flood of 1957 (VV.AA. 2000).

The Seminary-School of Corpus Christi was built with a rectangular plan of 170 x 74 span^2 with a single nave and two aisles on both sides (Espinosa and Rey 1590) and a the cruise, which gives it a Latin cross chapels.

Some years before the decision of Juan de Ribera to erect his Seminary-School, the city of Valencia bought a number of houses to proceed with the opening of the Main Square of the University.

Remember that on one hand the Patriarch exerts control tasks improvement of the University between 1572 and 1569 and, secondly, in these 70s up to eighteen churches of different religious congregations have been built.

In this period, the Patriarch Ribera will set his idea of building their own religious institution, mainly sponsored by the publication by Charles Borromeo of the *Instructionum Fabricae et Suppellectilis Ecclesiasticae* in 1577.

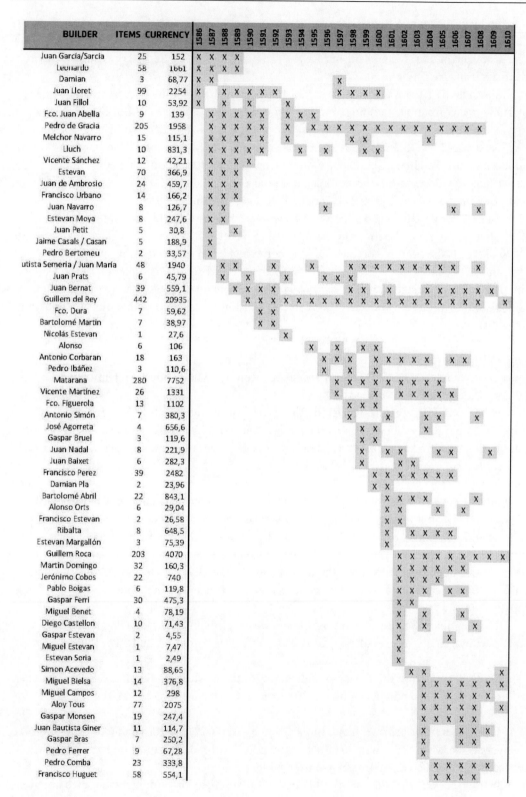

Figure 3. Builders who work al the Seminary-School of Corpus Christi of Valencia.

So, he decides to build a school-seminary to train priests close to the University (Casar 2001). The scope of the building was already clear before its construction and it was not improvising or expanding the surface the later years. This is reflected in the purchase of the previous houses, in 1580 Ribera starts to purchase from the north, which is the farthest area from the University (Lerma 2012).

Thanks to the documentation consulted on the Seminary-School, and from other authors (Gómez-Ferrer 1998; Ariciniega 2001) we have provided a better understanding of the other works in which the builders participated. Thus, we have recorded up to 247 people who worked on the construction of buildings in Valencia (Figure 3), whose characteristics are: (i) Trades: stonecutters, architects, painters, carpenters, masons, ringers, builders...

For the construction of the Seminary-School, the Patriarch Ribera was not enough to income from the miter (their properties), which were depleted and intended for peculiar objects and foundations that he was subsidized. So he had income only from his house and the generosity of the King, his friend Philip III (Cruilles and Monserrat 1876). He invested his fortune, inherited in 1571, after the death of his father, and he did not load the archbishops areas (Llopis 1997). Also he requested the patronage of the King by a letter sent in December 1594 and he was answered the same month, perceiving from the monarch 50,000 pounds (Boronat 1904), this is an amount of the 30% of the total investment in the Seminary-School. Thus, this is not a inconsiderable amount. The letter thanks Ribera the interest in to complete the mandate of the Tridentine Council and valued that the building would have borne by the archbishop, the Patriarch Ribera.

Conclusion

The architecture is a product of the era in which it is built. In this sense, the foundation, design and construction of the Seminary-School of Corpus Christi was the result of its contemporary events at the same end of the 16th century and early 17th century. The architecture of the Counter-Reformation is reflected in each of its parts; its proportions show the application of classic treaties and regulating lines. The purchase of the plots (previous houses in the area of the Seminary) shows that from the beginning Ribera knew the full size of the building.

Since the acquisitions were not performed at the same time, the construction of the Seminary was adapted. There is a correlation between the political, religious, social events, etc. that occurred during the construction of the building and the decisions made in the design and their implementation.

The existence of a strong influence of the work of Charles Borromeo *Instructiones Fabricae et Supellectilis Ecclesiasticae* is demonstrated. Virtually, all provisions raised by this Italian standard are met in our building.

References

Arciniega, L., (2001). *El monasterio de San Miguel de los Reyes*. Vol. I. Valencia: Generalitat Valenciana. ISBN vol I: 8448228782. ISBN complete work: 8448228774.

Boronat y Barrachina, P., (1904). *El B. Juan de Ribera y el R. Colegio de Corpus Christi, estudio histórico*. Valencia.

Casar, J. I. (2001) in *La Universitat i el seu entorn urbà*, 2001. ISBN: 8437051355.

Cruilles, V. C., Monserrat, M., (1876). *Guía urbana de Valencia: antigua y moderna*. Valencia.

Gallegos, M. E. (2004). http://www.sacredarchitecture.org/articles/ charles_borromeo_and_catholic_tradition/ Sacred *Architecture Journal*, Vol. 9.

Gómez-Ferrer, M., (1998). *Arquitectura en la Valencia del siglo XVI. El Hospital General y sus artífices. Valencia*: Albatros, 1998. ISBN: 8472742288.

Lerma, C. (2012). Análisis arquitectónico y constructivo del Real Colegio de Corpus Christi de Valencia. PhD thesis. http://hdl.handle.net/10251/ 18239.

Llopis, A., Perdigón, L., Taberner, F. (2004) *Valencia 138 a.C.-1929: De la fundación de la ciudad romana a la configuración y colmatación de la ciudad burguesa*. Faximil. ISBN 9788493339524.

Llopis Verdú, J. (2002a). Gaspar Gregori y la introducción de la metodología proyectual renacentista en Valencia. *Journal EGA,* nº 7 pp. 48-51. Las Palmas de Gran Canaria: Universidad LPGC. ISSN: 1133-6137.

Llopis Verdú, J. (1997). *Análisis de los órdenes clásicos en la arquitectura renacentista valenciana: el colegio de Corpus Christi*. Valencia: Universidad Politécnica de Valencia.

Llopis Verdú, J. (2002b). *Análisis gráfico de las formas clasicistas de la arquitectura valenciana*. IX Congreso Internacional de Expresión Gráfica Arquitectónica.

Vizcaíno Martí, Mª. E. (1999). Azulejería Barroca Valenciana. Valencia: Federico Doménech, 1999. ISBN: 8495031167.

VV.AA. (2000). Newspaper *Las Provincias*, book 2. Espinosa, M., Rey, G. (1590). Concierto entre D. Miguel de Espinosa y Guillem del Rey para la construcción de la Iglesia del Colegio de Corpus Christi.

VV. AA. (1999). Felipe III, poco Rey para tanto reino. *Journal La aventura de la Historia,* nº9. Newspaper *El mundo*.

In: Modeling Human Behavior: Individuals and Organizations   ISBN: 978-1-53610-197-3
Editors: L. Jódar Sánchez, E. de la Poza Plaza et al.   © 2017 Nova Science Publishers, Inc.

Chapter 12

MODELING OF HUMAN CAPITAL AND IMPACT ON EU REGIONAL COMPETITIVENESS

Lenka Fojtíková*, Michaela Staníčková and Lukáš Melecký
Department of European Integration, Faculty of Economics
VŠB - Technical University of Ostrava
Ostrava, Czech Republic

ABSTRACT

Human capital is considered to be an important factor of economic growth and development, as well as one of the sources of competitiveness and competitive advantages of individuals, companies, orgamizations and internationl integration groupings. In order to attain highly skilled human capital, economic entities should improve their labor market competitiveness and increase investments in education, science and technology. This chapter reconsiders influence of human capital on competitiveness of regional economy (regional competitiveness). Bearing in mind that competitiveness of regional economy has been dominantly defined by different factors of competitiveness, this research analyzes the role and the significance of human capital and focuses on different aspects of human capital in the light of specialized European Union measurement approach – the Regional Competitiveness Index and its dimensions referring to human capital, i.e., Health, Quality of Primary and Secondary Education and Higher Education/Training and Lifelong Learning, as well as their relative advantage over European Union. Europe's competitiveness depends on a multiplicity of actions that can optimise the potentials within its regions because regions are increasingly becoming the drivers of the economy. All regions possess different development opportunities – however, it does not mean they are competitive. To be competitive, regions have to use these options enough and effectively. The chapter is focused on using the Data Envelopment Analysis methodology for comparison the dynamic efficiency within the group of European NUTS 2 regions. DEA seems to be suitable toll for setting an effective/ineffective position of each region within the EU because measures numerical grades of efficiency of economical processes within evaluated regions. In the chapter, DEA method is applied to 268 NUTS 2 regions of 27 EU Member States and evaluate

---

* Sokolská třída 33, 701 21 Ostrava 1, Czech Republic; lenka.fojtikova@vsb.cz.

their efficiency within the selected factors of competitiveness based on the RCI 2010/2013, and recognize spatial variations in location factors influencing the attractiveness of regions. Results obtained by calculating the Malmquist Productivity Index indicate in which NUTS 2 region should be policy making authorities in order to stimulate regional development and provide more quality of life and well-being to the EU citizens.

Keywords: competitiveness, DEA, EU, human capital, malmquist productivity index, RCI

INTRODUCTION

Over the past half century, the European Union (EU) has been successful in securing high and rising living standards for their citizens. However, it is currently facing critical economic and social challenges. Despite past success, the financial and economic crisis of the last five years has led several European economies and the EU itself to one of their most difficult moments in the post-World War II period. The EU is going through one of the most difficult periods since its establishment, with multiple challenges facing the region's policy-makers. There is widespread agreement that the root causes of this prolonged crisis lie in the lack of competitiveness of many countries (WEF, 2015). The EU faces increased competition from other continents, their nations, regions and cities. Territorial potentials of European regions and their diversity are thus becoming increasingly important for the development of the European economy, especially now in times of globalisation processes in world economy. The EU, its regions and larger territories are increasingly affected by developments at the global level. New emerging challenges impact on territorial development and require policy responses. Territorial imbalances on the other hand challenge the economic, social and territorial cohesion within the EU. Contributions from cities, regions and larger territories are important for Europe's position in the world and thus for the achievement of the aims set out in European growth strategies aiming on competitiveness, i.e., the Lisbon strategy for period 2000-2010 and the Strategy Europe 2020 for period 2010-2020. These strategies were and still are aimed to make Europe the world's leading knowledge-economy, based on the principle of sustainable development. But actions are needed at all levels of government – European, national and regional/local levels – if these ambitions are to be realised. Europe's global competitiveness depends on a multiplicity of actions that can optimise the potentials within its regions, cities and rural areas.

The EU competitiveness depends on contributions from regions, cities and rural areas in all corners of the continent. An asset for Europe is its rich regional diversity which for each region and larger territory represents a unique set of potentials and challenges for development calling for a corresponding targeted policy mix to become reality. This regional diversity represented by specific territorial endowment is also possible to consider as a competitive advantage of each region. European policy development has thus moved towards recognising the territorial dimension in many policies and the added value from an integrated approach when searching for development opportunities. Modern strategic objectives for territories opt both for improving the cohesion and the competitiveness of the area, and to improve both the attractiveness for investments and the liveability for people. In doing so a number of territorial trends, perspectives, policy impacts and scenarios should be considered which influence policy aims of cohesion and balance and the competitiveness of territories. Opportunities and challenges of different territorial types such as regions, cities, rural areas and areas with specific characteristics and important themes as accessibility, innovation and hazards should be part of this. Territories are living legacies from the past and contain development potentials for the future. Trends and perspectives can be identified, and the impacts of policies can be seen. The interplay of all these factors underpins a territory's demographic, economic, social, cultural and ecological development dynamics. Thus each territory, be it continent, region, metropolitan area or village, has its own unique settings and development conditions. Knowledge and understanding of the territory is an important prerequisite for ensuring a future development for competitive attractive and liveable places.

Increasing the competitiveness of Europe and its regions is one of the main aims of the EU. This involves focusing on growth and jobs, as well as growing the necessary preconditions for the future mainly in terms of a Knowledge Based Economy and Information Society. The creation and development of the knowledge based society and knowledge economy are perceived as one of the most important priorities of the modern society and its lifestyle development, as well as of social, economic, political development, science and technological progress (Melnikas, 2011). There is a direct and very high correlation between competitiveness of an economy and its potential to grow: the more competitive economy of a country is; the higher growth rates it achieves. This goes for the other way round as well, the more dynamic growth of an economy, the bigger chances of achieving a higher level of competitiveness. Only a certain and this type of territories appears to be really successful with regard to the EU strategies. However, there are also examples of other types of areas which are performing well with regard to economic development. The key to success seems mainly to lie in the active use of territorial potentials for the development of economic functions across a wider area, and support through national policies. In short, territories have diverse potentials and challenges. Territories entail the long term structures that shape living and working conditions now and for future generations. Territories matter for the competitiveness and cohesion of Europe, for sustainable development and for European citizens and businesses (ESPON, 2006a). All territories possess development opportunities. However, to make sound policy decisions requires evidence, knowledge and understanding of the position of regions and cities both within Europe, and also globally.

In the EU, the process of achieving an increasing trend of performance and a higher level of competitiveness is significantly difficult by the heterogeneity of countries and regions in many areas. Although the EU is one of the most developed parts of the world with high living standards, there exist significant and huge economic, social and territorial disparities having a negative impact on the balanced development across Member States and their regions, and thus weaken EU's performance in a global context. The history of European integration process in the past five decades was and is thus guided by striving for two different objectives: to foster economic competitive-ness and to reduce differences (Molle, 2007). The support of cohesion and balanced development together with increasing level of EU competitiveness thus belong to the temporary EU's key development objectives. In relation to competitiveness, performance and efficiency are complementary objectives, which determine the long-term development of countries and regions. Measurement, analysis and evaluation of productivity changes, efficiency and level of competitiveness are controversial topics acquire great interest among researchers; see e.g., (Camanho and Dyson, 2006; Khan and Soverall, 2007). Motivation of this paper is based on mutual relationship between two significant themes presented by efficiency and competitiveness in the context of national economies. At a time when the EU and all its Member States have to deal with increased pressures on public balances, stemming from demographic trends and globalisation, the improvement of the efficiency of public spending features high on the political agenda. This fact is closely connecting with the aim of competitiveness, because rational using of sources/funds for activities could ensure the effective provision of these activities and their corresponding results, which is having an impact on the competitive advantages of each territory. From this point of view, the main aim of the paper is to measure efficiency changes over the reference period and to analyse a level of competitive performance in individual EU NUTS 2 regions based on advanced Data Envelopment Analysis (DEA) approach – the Malmquist Productivity Index (MPI) measuring the change of technical efficiency and the change of technological efficiency. Because efficiency analysis is closely lined with competitiveness, the EU Regional Competitiveness Index (RCI) is used as initial database and approach. The paper shows that the efficiency in particular varies significantly between evaluated NUTS 2 regions, resp. where is potential for increased efficiency and improved thus competitiveness. The main focus is thus to evaluate the RCI time series which may serve as a tool to assist the EU NUTS 2 regions in setting the right priorities to further increase their competitiveness. Regions have indeed to pick priorities for their development strategies. The economic crisis made this even more difficult as public funding becomes scarcer. Efficiency analysis of competitiveness could provide a guide to what each region should focus on, taking into account its specific situation and its overall level of development.

THEORETICAL BACKGROUND

Human Capital and Competitiveness

Today more then ever, there is a strong incentive for individuals to continuously improve themselves professionally. The emphasis on education and development of professional skills is being made as a condition of successful competition in national and global labor market. Thus, governments should create an environment favorable for human capital creation that will, in return, benefit both the country as a whole, and its individuals. Otherwise, individuals will seek better opportunities elsewhere (Matovac, Bilas and Franc, 2010).

But what does human caopital mean? Human capital is a broad concept encompassing many different types of investment in people. In short, it can be defined as the abilities, knowledge and skills embodied in people and acquired through education, training and experience. Knowledge, skills, creativity, innovativeness, ability to learn and other valuable features people own have become a key element in modern economy, both for their earning capacity and competitiveness and other economic performances of a company as well. Terms 'IT society,' 'Learning society,' 'Network economy,' 'New economics,' 'Knowledge based economy,' 'Knowledge economy,' 'Innovative economy' have been used to describe growing importance of intellectual capital on competitiveness of a company and economic and social development of a country (Djurica, Djurica and Janicic, 2014).

The importance of human capital is being recognized in both developed and developing countries considering that we live in the era of globalization, fierce competition, continuous technology development and innovation. It is commonly accepted that human capital accumulation induces various externalities, especially in the area of technology and innovation. Human capital is considered to be a crucial input for the development of new technologies and a necessary factor for their adoption and efficient use. The three basic conclusions emerging from the large body of empirical work about the consequences of formal education on labor markets are that: higher levels of education are accompanied by higher wages, lower unemployment probabilities, and higher labor force participation rates. Most of the work has been done on the link between schooling and wages. This is because the resulting wage increase is the most important economic consequence of higher levels of formal education (De la Fuente and Ciccone, 2002). Overall, investing in the development of high quality human capital is expected to have a positive impact on employment and economic growth.

Two centuries ago, Adam Smith in his well-known book, The Wealth of Nations, underlined that improvement of workers' skills was a fundamental source of economic progress. Frank Knight has also emphasized in his papers published in the '30s, that enhancement of intellectual capital could compensate for the law on decreasing labor and capital returns. In 1962, Denison published results of empiric research on the US economic growth resources for the period between 1909 and 1958, concluding that knowledge, skills and workers' energy were key determinants of economic growth at that time. The latest economic theories, developed in the course of mid-eighties by Paul M. Romer, Robert E. Lucas, Robert Baro and others, known as endogenous growth theories, have been founded on a postulate that investment in human capital, innovations in knowledge significantly contributes to the economic growth. These theories require that growth models should simultaneously analyze both tangible and intangible factors and their mutual relations. The latest papers in marketing and management areas underline that intellectual capital is the main source of value creation and competitiveness (Djurica, Djurica and Janicic, 2014).

At the beginning of the 21st century, the gap in living standards between rich and poor nations is large and rising. It is generally accepted that deficiency in human capital is an important factor, i.e., obstacle to country's growth. If a country lacks human capital of good quality, then it has fewer opportunities for growth and development. The rapid structural change caused by globalization and technological change has increased the importance of human capital over the past years. In the rich economies, this structural change increased the pressure on the suppliers of the less qualified labor force. Physical work is substituted by machines at home and by cheaper labor input from abroad. As a reaction, rich, more developed countries can either shield themselves from globalization, which would be negative for prosperity, cut the wages of less qualified workers, accept higher unemployment, or they can raise the skills of their workers (Deutche Bank Research, 2005).

Over the past few decades social and economic picture of the world has changed and resulted in increased demanding on human capital of individuals worldwide. The collapse of socialist regimes in Eastern Europe, differences between incomes in developed and less developed countries, other labor market characteristics, wars, terrorism and human rights violation are just some of the factors that led to increased changed aspects of human capital and its flows. Worldwide globalization process has resulted in a change of the human capital experience and labor market characteristics of many countries. The importance of human capital is recognized in both developed and developing countries considering that we live in the era of globalization, fierce competition, continuous technology development and innovation. In knowledge based economy, which would be the 21st century's trademark, contribution of intellectual capital to growth and development of an enterprise and general economic growth of every country has become much greater. Accordingly, economic theory focus has lately been re-orientated from material to nonmaterial resources, i.e., from tangible to intangible factors pertaining to company's competitiveness and national economy as a whole, with knowledge management; i.e., intellectual capital management as a core research subject. Despite the fact that knowledge and other elements of human, i.e., intellectual capital represent key items of nonmaterial assets and underpinning power of long-term sustainable competitive advantage of a company, employees in a number of companies, and even human resources managers, have not been sufficiently aware of strategic importance of human resources management (Djurica, Djurica and Janicic, 2014).

Efficiency and Competitiveness

In recent years, the topics about measuring and evaluating of competitiveness and efficiency have enjoyed economic interest. These multidimensional concepts remain ones of the basic standards of performance evaluation and it is also seen as a reflection of success of area in a wider comparison. Efficiency and competitiveness are thus complementary objectives, which determine the long-term performance development of area. The exact definition of competitiveness is difficult because of the lack of mainstream view for understanding this term. Competitiveness remains a concept that can be understood in different ways and levels despite widespread acceptance of its importance. The concept of competitiveness is distinguished at different levels – microeconomic, macroeconomic and regional. Anyway, there are some differences between these three approaches; see e.g., (Krugman, 1994). There are differences not only among concepts of competitiveness, but also different approaches to measurement and evaluation of competitiveness exist; see e.g., (Fojtíková et al, 2014). Competitiveness is monitored characteristic of national economies which is increasingly appearing in evaluating their performance and prosperity, welfare and living standards. The need for a theoretical definition of competitiveness at macroeconomic level emerged with the development of globalization process in the world economy as a result of increased competition between countries. Despite that, growth competitiveness of the territory belongs to the main priorities of countries' economic policies. In last few years the topic about regional competitiveness stands in the front of economic interest. The concept of competitiveness has quickly spread into regional level, but the notion of regional competitiveness is also contentious. In the global economy regions are increasingly becoming the drivers of the economy and generally one of the most striking features of regional economies is the presence of clusters, or geographic concentrations of linked industries (Porter, 2003). Current economic fundamentals are threatened by shifting of production activities to places with better conditions. Regional competitiveness is also affected by the regionalization of public policy because of shifting of decision-making and coordination of activities at regional level. Within governmental circles, interest has grown in the regional foundations of national competitiveness, and with developing new forms of regionally based policy interventions to help improve competitiveness of every region and major city, and hence the national economy as a whole. Regions thus play an increasingly important role in the economic development of states (Staníčková and Skokan, 2013). Nowadays competitiveness is one of the fundamental criteria for evaluating economic performance and reflects the success in the broader comparison. Territories need highly performing units in order to meet their goals, to deliver the products and services they specialized in, and finally to achieve competitive advantage. Low performance and not achieving the goals might be experienced as dissatisfying or even as a failure. Moreover, performance, if it is recognized by others organizations, is often rewarded by benefits, e.g., better market position, higher competitive advantages, financial condition etc. Performance is a major prerequisite for future economic and social development and success in the broader comparison. Differences in productivity performance across territories are seen by government as important policy targets. For a number of years, government objectives have been set not only in terms of improving national productivity performance against other countries but also in creating conditions to allow less productive countries to reduce the 'gap' between themselves and the most productive ones.

Comparative analysis of efficiency in public sector is thus starting point for studying the role of efficiency, effectiveness and performance regarding economic governance of resources utilization by public management for achieving medium/long-term objectives of economic recovery and sustainable development of national economies (Mihaiu, Opreana and Cristescu, 2010). Increasing productivity is generally considered to be the only sustainable way of improving living standards in the long term. Statistical evidence to help policy makers understand the routes to productivity growth, especially those which can be influenced by government, can help lead to better policy. Efficiency is thus a central issue in analyses of economic growth, the effects of fiscal policies, the pricing of capital assets, the level of investments, the technology changes and production technology, and other economic topics and indicators. The efficiency can be achieved under the conditions of maximizing the results of an action in relation to the resources used, and it is calculated by comparing the effects obtained in their efforts. In a competitive economy, therefore, the issue of efficiency can be resolved by comparing these economic issues. The efficiency is provided by the relationship between the effects, or outputs such as found in literature review, and efforts or inputs. The relationship is apparently simple, but practice often proves the contrary, because identifying and measuring inputs and outputs in the public sector is generally a difficult operation. Figure 1 illustrates the conceptual framework of efficiency and effectiveness. The efficiency is given by the ratio of inputs to outputs, but there is difference between technical efficiency and allocative efficiency. Technical efficiency implies a relation between inputs and outputs on the frontier production curve, but not any form of technical efficiency makes sense in economic terms, and this deficiency is captured through allocative efficiency that requires a cost/benefit ratio. The effectiveness implies a relationship between outputs and outcomes. In this sense, the distinction between output and outcome must be made. The outcome is often linked to welfare or growth objectives and therefore may be influenced by multiple factors. The effectiveness is more difficult to assess than efficiency, since the outcome is influenced political choice.

Based on information mentioned above, the concept of competitiveness is usually linked to productivity (Porter, 1990). Competitiveness may be defined as a measure of the degree in which each economic entity can compete with economic entity. However, the concept of competitiveness may be applicable not only to firms, but also to whole economies. An economy is competitive if firms in that economy face lower unit costs than firms from other economies. Every factor that increases the productivity and, therefore, lowers the unit costs of firms in an economy contributes to the competitiveness of the respective economy (Charles and Zegarra, 2014).

According to the Institute for Management and Development (2012), competitiveness is "a field of economic knowledge, which analyses the facts and policies that shape the ability of a nation to create and maintain an environment that sustains more value creation for its enterprises and more prosperity for its people" (p. 502). In other words, competitiveness measures "how a nation manages the totality of its resources and competencies to increase the prosperity of its people" (IMD, 2012, p. 502). This understanding of competitiveness and interpretation of this concept is thus very closely linked with understanding of efficiency and effectiveness concepts, see Figure 1.

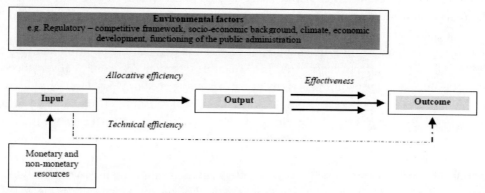

Source: Mandl, Dierx and Ilzkovitz (2008); Own elaboration, 2016.

Figure 1. Relationship between Efficiency and Effectiveness.

MATERIALS AND METHODS

Regional Competitivenes Index Background

The efficiency analysis starts from building database of indicators that are part of Regional Competitiveness Index (RCI) approach created by Annoni and Kozovska (2010) in 2010, and then in 2013 updated by Annoni and Dijkstra (2013). The roots of the RCI lay in the most known competitiveness indicator, the Global Competitiveness Index (GCI) reported by the World Economic Forum (WEF). To improve the understanding of territorial competitiveness at the regional level, the European Commission has developed this index which shows the strengths and weaknesses of each of the EU NUTS 2 regions. It covers a wide range of issues related to territorial competitiveness including innovation, quality of institutions, infrastructure (including digital networks) and measures of health and human capital, see Table 1.

The RCI is based on eleven pillars describing both inputs and outputs of territorial competitiveness, grouped into three sets describing basic, efficiency and innovative factors of competitiveness. Basic pillars represent the basic drivers of all economies. They include (1) Quality of Institutions, (2) Macro-economic Stability, (3) Infrastructure, (4) Health and the (5) Quality of Primary and Secondary Education. These pillars are most important for less developed regions and these five pillars are taken to represent the key basic drivers of all types of economies. The efficiency pillars are (6) Higher Education and Lifelong Learning (7) Labour Market Efficiency and (8) Market Size. As a regional economy develops, other factors related to a more skilled labour force and a more efficient labour market enter into play for its advancement in competitiveness and are part of the Efficiency group. At the most advanced stage of development of a regional economy, drivers of improvement are part of the Innovation group which consists of three pillars: (9) Technological Readiness, (10) Business Sophistication and (11) Innovation. This group plays a more important role for intermediate and especially for highly developed regions. Overall, the RCI framework is designed to capture short- as well as long-term capabilities of the regions.

Table 1. The RCI Framework – Sub-indices and Relevant Pillars

Basic Group	Efficiency Group	Innovation Group
Institutions	Market Size	Technological Readiness
Macroeconomic Stability	Labour Market Efficiency	Business Sophistication
Infrasructure	Higher Education/Training and Lifelong Learning	Innovation
Health		
Quality of Primary and Secondary Education		

Source: Annoni and Kozovska, 2010; Own elaboration, 2016.

Within the RCI pillars, there are three pillars having basic line and impact with human capital dimension of competitiveness, i.e., Health, Quality of Primary and Secondary Education and Higher Education/Training and Lifelong Learning. Health and basic education: health of workforce and basic education received by the population are clearly key aspects of a productive and efficient economy. This pillar aims to measure the incidence of major invalidating illnesses, infant mortality, life expectancy and the quality of primary education. Higher education and training: if basic education is the starting point of a ductile and efficient workforce, higher education and continuous training are crucial for economies not restricted to basic process and products. This pillar describes secondary and tertiary education together with the extent of staff training.

Pillars of the RCI are grouped according to the different dimensions (input versus output aspects) of regional competitiveness they describe. The terms "inputs" and "outputs" are meant to classify pillars into those which describe driving forces of competitiveness, also in terms of long-term potentiality, and those which are direct or indirect outcomes of a competitive society and economy, defined by Annoni and Kozovska (2010). From this point of view, the RCI approach seems to be convenient with respect to used methodology of DEA and its division to input and output nature of incoming database. The RCI data file consists of 66 indicators in 2010, and 73 indicators in 2013; but initial indicators are not used in the paper. Database of analysis is created by three sub-indices of the RCI, i.e., SubInd1: the RCI-Basic, SubInd2: the RCI-Efficiency and SubInd3: the RCI-Innovation. Within these three sub-indices (separately for the RCI 2010 and the RCI 2013), eleven pillars of RCI are included and these represent databse of analysis. These groups are purportedly linked: a good performer in the Innovation group is expected to also be a good performer in the Efficiency and the Basic groups as they are instrumental to increasing levels of competitiveness. As regions move along the path of development, their socio-economic conditions change and different determinants become more important for the regional level of competitiveness. As a result, the best way to improve the competitiveness of more developed regions will not necessarily coincide with the way to improve less developed regions. To take this into account, and following the WEF-GCI approach, the EU NUTS 2 regions are divided into "medium," "intermediate" and "high" stages of development in the RCI 2010 (Dijsktra, Annoni and Kozovska, 2011). This is done on the basis of regional GDP per head for 2007 in PPS (purchasing power standard). The EU NUTS 2 regions are classified into the three stages of development in the RCI 2010, according to the thresholds listed in Table 2. On the basis of membership of each region to suitable stage of development, this region within each sub-pillar is assigned by weight. In the RCI 2013, the weighting system and regions classification into development stages have been slightly modified, also following the suggestions by the WEF team in charge of the GCI. Five classes, instead of three of the RCI 2010, are used to allow for a smoother change in the weighting values across development stages based on GDP per head for average 2007-2008-2009 in PPS. In fact, the RCI does not have any transition stages which are instead used in WEF-GCI with country specific set of weights, but by adding two more classes, the RCI 2013 try to cope with this issue, see Table 3.

Table 2. The RCI 2010 Weighting Scheme

Percentage of GDP (PPS/inhabitant)	Development stage	Weights (w)			SUM
		SubInd1: the RCI-Basic	SubInd2: the RCI-Efficiency	SubInd3: the RCI-Innovation	
< 75	Medium (M)	0.40000	0.50000	0.1000	100%
≥ 75 and < 100	Intermediate (I)	0.30000	0.50000	0.2000	100%
≥ 100	High (H)	0.20000	0.50000	0.3000	100%

Source: Annoni and Kozovska, 2010; Own elaboration, 2016.

Table 3. The RCI 2013 Weighting Scheme

Percentage of GDP (PPS/inhabitant)	Development stage	Weights (w)			SUM
		SubInd1: the RCI-Basic	SubInd2: the RCI-Efficiency	SubInd3: the RCI-Innovation	
< 50	1	0.35000	0.50000	0.15000	100%
≥ 50 and < 75	2	0.31250	0.50000	0.18750	100%
≥ 75 and < 90	3	0.27500	0.50000	0.22500	100%
≥ 90 and < 110	4	0.23750	0.50000	0.26250	100%
≥ 110	5	0.20000	0.50000	0.30000	100%

Source: Annoni and Dijkstra, 2013; Own elaboration, 2016.

Data Envelopment Analysis Background

Data Envelopment Analysis (DEA) was first proposed by A. Charnes, W. W. Cooper and E. Rhodes (CCR model) in 1978 (Charnes, Cooper and Rhodes, 1978). Since DEA was first introduced, researchers in a number of fields have quickly recognized that it is an excellent and easily used methodology for modelling operational processes for performance evaluations, e.g., (Hančlová, 2013; Melecký, 2013a, b; Staníčková, 2013). This has been accompanied by other developments. In DEA, there are several methods for measuring efficiency, besides the basic DEA models, certain modifications exist. DEA is mathematical approach for providing a relative efficiency assessment and evaluating performance of a set of peer entities called Decision Making Units (DMUs) which convert multiple inputs into multiple outputs. DEA is thus a multicriteria decision making method for evaluating efficiency and productivity of a homogenous group (DMUs). Definition of a DMU is generic and flexible. DEA is convenient to determine the efficiency of DMU, which are mutually comparable – using the same inputs, producing the same outputs, but their performances are different. In recent years, research effort has focused on investigation of the causes of productivity change and its decomposition. Malmquist Productivity Index (MPI) has become the standard approach in productivity measurement over time within the non-parametric research. The MPI has been introduced firstly by Caves, Christensen and Diewert in 1982 (Caves, Christensen and Diewert, 1982). Färe et al. (1994 a, b) defined an input-oriented productivity index as the geometric mean of the two MPIs developed by Caves et al. Although it was developed in a consumer context, the MPI recently has enjoyed widespread use in a production context, also in territorial analysis. The MPI can be used to construct indexes of input, output or productivity, as ratio of input or output distance functions. There are various methods for measuring distance functions, and the most famous one is the linear programming method. The MPI allows measuring of total productivity by means of distance-functions calculation, which can be estimated by solution of mathematical programming problems of DEA kind.

Suppose there are n DMUs which consume m inputs to produce s outputs. If a performance measure (input/output) is added or deleted from consideration, it will influence the relative efficiencies. Empirically, when the number of performance measures is high in comparison with the number of DMUs, then most of DMUs are evaluated efficient. Hence, the obtained results are not reliable. There is a rough rule of thumb (Cooper, Seiford and Tone, 2007) which expresses the relation between the number of DMUs and the number of performance measures as follows (1):

$$n \geq \max\{3(m+s), m \times s\} \tag{1}$$

Nevertheless, in some applications the number of performance measures and DMUs do not meet the mentioned formula (1). To tackle this issue, it should select some performance measures in a manner which comply (1) and impose progressive effect on the efficiency scores. These selected inputs and outputs calls selective measures. But formula (1) needs more considerations. Toloo et al. checked more than 40 papers that contain practical applications and statistically, they found out that in nearly all of the cases the number of inputs and outputs do not exceed 6 (Toloo, 2012). A simple calculation shows that when $m \leq 6$ and $s \leq 6$, then $3(m+s) \geq m \times s$. As a result, in this paper instead of using (1), following formula (2) is applied:

$$n \geq 3(m+s) \tag{2}$$

In the case of this paper, the rule of thumb is met, because number of DMUs is three times higher than sum of input and outputs, i.e., $212 \geq 3 (7 + 4)$, $268 \geq 3 (11)$, $268 \geq 33$ (for EU15 NUTS 2 regions), and also $56 \geq 3 (7 + 4)$, $56 \geq 3 (11)$, $56 \geq 33$ (for EU12 NUTS 2 regions).

Suppose there are n DMUs (DMU_j $j=1,...,n$) with m inputs, $\mathbf{x}_j = (x_{1j},...,x_{mj})$, and s outputs, $\mathbf{y}_j = (y_{1j},...,y_{sj})$. The CCR model (3) measures the efficiency of the under evaluation DMU, i.e., DMU_o for $o \in \{1,...,n\}$:

$$\text{maximize } \theta_o = \sum_{r=1}^{s} u_r y_{ro},$$

subject to

$$\sum_{i=1}^{m} v_i x_{io} = 1,$$

$$\sum_{r=1}^{s} u_r y_{rj} - \sum_{i=1}^{m} v_i x_{ij} \leq 0, \quad j=1,...,n, \tag{3}$$

$$u_r \geq 0, \quad \forall r,$$

$$u_i \geq 0, \quad \forall i,$$

where v_i and u_r are the unknown i^{th} input and o^{th} output weights. It is proved that the CCR model is always feasible and its optimal objective value is bounded; more precisely $0 < \theta_j^* \leq 1$ for $j=1,...,n$.

In contrast to traditional DEA models which measure the efficiency of a DMU, the MI enables to measure the productivity change of a DMU between two time periods, t and $t + 1$. The MI is defined as the product of Catch-up and Frontier-shift terms. The catch-up term deals with the degree to which a DMU improves or worsens its efficiency – technical efficiency change, while the frontier-shift term shows the change in the efficient frontiers between the two time periods – technological efficiency change. It is denoted $DMU_j^1 = (\mathbf{x}_j^1, \mathbf{y}_j^1)$ and $DMU_j^2 = (\mathbf{x}_j^2, \mathbf{y}_j^2)$ to show the data set of DMU_j for Period 1 and Period 2. Clearly, there are two efficient frontiers with these assumptions. The catch-up effect from Period 1 to Period 2 is defined as follow (4):

$$\text{Catch-up} = \frac{\text{Efficiency of }(\mathbf{x}_o^2, \mathbf{y}_o^2)\text{ with respect to Period 2 frontier}}{\text{Efficiency of }(\mathbf{x}_o^1, \mathbf{y}_o^1)\text{ with respect to Period 1 frontier}}. \tag{4}$$

There cases may be arisen:

Case No. 1: (Catch-up) < 1 shows regress in relative efficiency from Period 1 to Period 2.
Case No. 2: (Catch-up) = 1 indicates there is no change in relative efficiency.
Case No. 3: (Catch-up) > 1 displays progress in relative efficiency.
To evaluate the Frontier-shift effect more computations are required. Let have (5):

$$\phi_1 = \frac{\text{Efficiency of }(\mathbf{x}_o^1, \mathbf{y}_o^1)\text{ with respect to Period 1 frontier}}{\text{Efficiency of }(\mathbf{x}_o^1, \mathbf{y}_o^1)\text{ with respect to Period 2 frontier}}. \tag{5}$$

And also, let have (6):

$$\phi_2 = \frac{\text{Efficiency of } (x_o^2, y_o^2) \text{ with respect to Period 1 frontier}}{\text{Efficiency of } (x_o^2, y_o^2) \text{ with respect to Period 2 frontier}}. \qquad (6)$$

Using these notations, Frontier-shift effect can be defined as follows (7):

$$\text{Frontier-shift} = \sqrt{\phi_1 \phi_2} \qquad (7)$$

There are three possible cases:

Case No. 1: (Frontier-shift) < 1 shows regress in the frontier technology around DMUo.
Case No. 2: (Frontier-shift) = 1 indicates no changes in the frontier.
Case No. 3: (Frontier-shift) > 1 displays progress in the frontier technology.
Finally, the MPI is calculated as the product of (Catch-up) and (Frontier-shift) via (8):

$$\text{MI} = \text{Catch-up} \times \text{Frontier-shift} \qquad (8)$$

As a result, the MPI < 1 indicates deterioration in the total factor productivity of the DMU_o from Period 1 to Period 2; result of the MPI =1 shows there is no change in total factor productivity and the MPI > 1 shows progress in the total factor productivity (for more details see (Cooper, Seiford and Tone, 2007) and Table 4, where characteristics and trends of the MPI and efficiency change are shown).

Table 4. Characteristics and Trends of the MPI and Efficiency Change

Malmquist Productivity Index	Productivity	Malmquist Productivity Index Dimensions	Catch-up Frontier-shift	Technical Efficiency Change Technological Efficiency Change
> 1	Improving	MPI	< 1	Declining
= 1	Unchanging	Catch-up x Frontier-shift	= 1	Unchanging
< 1	Declining		> 1	Improving

Source: Own elaboration, 2016.

The following linear programming (9) measures the efficiency score of (x_o^s, y_o^s) with respect to Period t where $s, t \in \{1, 2\}$:

$$\text{maximize } \delta_s^t = \sum_{r=1}^{s} u_r y_{ro}^s,$$

subject to

$$\sum_{i=1}^{m} v_i x_{io}^s = 1,$$

$$\sum_{r=1}^{s} u_r y_{rj}^t - \sum_{i=1}^{m} v_i x_{ij}^t \leq 0, \quad j = 1, \ldots, n, \qquad (9)$$

$$u_r \geq 0, \quad \forall r,$$

$$u_i \geq 0, \quad \forall i.$$

In this model, $\frac{\delta_2^2}{\delta_1^1}$ indicates Catch-up effect, $\sqrt{\frac{\delta_1^1}{\delta_1^2} \times \frac{\delta_2^1}{\delta_2^2}}$ shows Frontier-shift effect and $\sqrt{\frac{\delta_1^2}{\delta_1^2} \times \frac{\delta_2^1}{\delta_2^1}}$ displays the MPI.

EMPIRICAL ANALYIS AND DISCUSSION

In the first step of analysis, the correction of database is made, i.e., computing the scores for all eleven pillars in both dimensions. Because of DEA requirement on positive values, it was necessary to correct the initial scores of eleven pillars for the RCI 2010 and the RCI 2013 (several regions showed negative values in some scores). For all EU NUTS 2 regions across two period 2010 and 2013, the correction was made as follows: it was calculated the minimum value for the whole group of eleven pillars, where min in the RCI 2010 equals -4.32 (for EU15 NUTS 2 regions) and -3.32 (for EU12 NUTS 2 regions); and the minimum value for the whole group of eleven pillars, where min in the RCI 2013 equals -3.19 (for EU15 NUTS 2 regions) and -2.98 (for EU12 NUTS 2 regions). Based on these minimum values, the following values were was added to the initial scores of calculated pillars, i.e., 5.00 (the RCI2010 for EU15 NUTS 2 regions), 4.00 (the RCI2010 for EU12 NUTS 2 regions), 4.00 (the RCI2013 for EU15 NUTS 2 regions) and 3.00 (the RCI2013 for EU12 NUTS 2 regions). All pillars gained positive values by this correction, what is required for DEA and this database thus coming into analysis. As it was already mentioned above, the RCI is composed of seven pillars of inputs and four pillars of outputs and these are thus compared in both reference years. Table 5 shows basic scheme for efficiency analysis.

Territorial background of analysis is applied at NUTS 2 region level within the group of EU27 Member States – 15 countries are classified as old EU Member States – origin countries from 1957 and countries joined to the European Communities in 1973, 1981, 1986 and 1995; and 12 countries belong to group of new EU Member States joined to the EU in 2004 and 2007. Croatia, joined to the EU as the 28th EU Member State in 2013, is excluded from analysis because of the MPI requirement, resp. Croatian NUTS 2 regions were not included in the RCI 2010 approach, but they were included in the RCI 2013 approach. There are 268 evaluated NUTS 2 regions, i.e., 212 NUTS 2 regions in EU15 countries and 56 NUTS 2 regions in EU12 countries.

Table 5. DEA Model with 7 Inputs and 4 Outputs

Inputs		Outputs
Institutions		Labor Market Efficiency
Macroeconomic stability		Market Size
Infrastructure	NUTS 2 Region	
Health *(Human Capital)*		
Quality of Primary and Secondary Education *(Human Capital)*		Business Sophistication
Higher Education/Training and Lifelong Learning *(Human Capital)*		Innovation
Technological Readiness		

Source: Own elaboration, 2016.

Many European NUTS 2 regions have strong economies, are well integrated into international networks and are the locus of enterprises and labour forces that are globally competitive. However, not all regions make a strong contribution towards competitive aims. How well are the EU's regions performing, and what makes a region competitive? Europe is a continent of special places - historic cities, coasts and cultural landscapes so familiar that it is easy to overlook their global significance (ESPON, 2006b). In order to get an impression on where EU NUTS 2 regions are placed in terms of their stage of development is used as a reference the RCI thresholds for classifying European regions on the basis of their stage of development. Within efficiency analysis of competitiveness, comparison of the RCI 2013 with the RCI 2010 and recognition of development in efficiency competitiveness is made, results provide evidence that territorial capital and potentials for development are inherent in the regional diversity that is a major characteristic of Europe. Results of efficiency analysis by advanced DEA approach – the MPI are presented in Tables 6 – 8. Dimensions of the MPI, i.e., Catch-up and Frontier-shift are shown in Figures 2 – 3. In following Tables 6 – 8, results of the MPI and its dimensions are highlighted by the traffic light method. Range of colours of this method changes from green colour, through shades of yellow colour to red colour (in black and white printed version – range of grey colour). Regions with the highest and higher values of the MPI, Catch-up and Frontier-shift mean better level of efficiency and thus competitiveness are highlighted by green colour – the higher value, the darker shadow of green colour (in black and white printed version – highlighting by grey colour – the higher value, the darker shadow of grey colour). On the contrary, regions with the lowest and lower values of the MPI and its two dimensions (Catch-up and Frontier-shift) mean worse level of efficiency, resp. level of inefficiency are highlighted by red colour – the lower value, the darker shadow of red colour (in black and white printed version – highlighting by grey colour – the lower value, the lighter shadow of grey colour; in the case of zero values – the darker shadow of grey colour). Regions with values of the MPI and its dimension between group of efficiency (shadows of green colour) and inefficiency regions (shadows of red colour) are highlighted by shadows of yellow colour (in black and white printed version – highlighting by white colour).

Before discussion of efficiency analysis results, it is necessary to remind that with respect to initial database, the RCI index, its sub-indices and scores are computed on the basis of the 11 pillar scores presented in Table 1, and the three sub-indices – basic, efficiency and innovation – are computed as simple arithmetic averages of the three groups of pillars. It is possible to state an increasing heterogeneity in the performance of regions across the three pillar groups with more regions having similar scores in the Basic pillar group, as expected, with the exception of some of the newest EU Member States. Performance in the Innovation pillar group shows highest diversity across regions, suggesting the different levels of sophistication of regional economies. The Basic group of pillars consists of Institutions, Macroeconomic Stability, Infrastructure, Health and Basic Education pillars. They are considered as factors which are strictly necessary for the basic functioning of any economy and cover aspects like unskilled or low skilled labour force, infrastructures, quality of governance and public health, which are also important economic and social determinants. The second group of pillars includes Higher Education and Lifelong Learning, Labour Market Efficiency and Market Size. They describe a socio-economic environment more developed than the previous one, with a potential skilled labour force and a more structured labour market. The last group of pillars comprises all the high tech and innovation related pillars:

Technological Readiness, Business Sophistication and Innovation. A region scoring high in these aspects is expected to have the most competitive economy. The three sub-indices were then aggregated using a weighted linear function with weights depending on the stage of development for each region as discussed in Table 2 and Table 3. In the RCI 2013 approach, higher number of development stage classes was adopted – 5 instead of 3, to allow for a smoother weight change across different stages and slightly increase of the weights assigned to the innovation pillar group for the lowest developed regional economies to reward innovative policies even in lagging behind regions, as also recommended by the WEF-GCI team (Annoni and Dijsktra, 2013).

Consistent with the theory of economic growth and economic development, the RCI results confirm that the most competitive regions are those with the highest level of economic development. In the field of competitiveness, the best "top-ten" group includes Utrecht, the highest competitive region in both editions of the RCI, the London area and the area including Oxford, the two Netherland regions of Noord and Zuid Holland which comprise Amsterdam, the Danish region Hovedstaden (including Copenhagen), Stockholm and Île de France (including Paris). The other entries in the top-ten are the Frankfurt region (Darmstadt) and the Surrey, East and West Sussex in the United Kingdom. It is striking that seven out of the top-ten are either capital regions or regions including large cities. At the other end of the competitiveness scale, it is possible to find some regions which are unfortunately steadily worst performers. These are the Bulgarian region Severozapaden, the Greek region Notio Aigaio, and two southern Romanian regions Sud-Est and Sud-Vest Oltenia. The RCIs show a more polycentric pattern with strong capital and metropolitan regions in many parts of the EU. Some capital regions are surrounded by similarly competitive regions, but in many countries the regions neighbouring the capital are less competitive. As this was also observed for the both RCI editions, the RCIs show that in the past three years (2013 to 2010 edition) no spill over effects helped to lift these lagging-behind surrounding regions. The general economic and financial crisis certainly did not help. Thus, the substantial disparities within several countries also highlight the need for regional analysis and the limits of a purely national approach. The RCI 2010 and the RCI 2013 results underline that competitiveness has a strong regional dimension, which national level analysis does not capture (Annoni and Dijsktra, 2013). For example, in some countries like France, Spain, United Kingdom, Slovakia, Romania, Sweden and Greece, the level of variability of the RCI scores is particularly high with the capital region almost always being the best performer within the country. Italy is an exception as Lombardy is the Italian most competitive region. These results demonstrate that territorial competitiveness in the EU has a strong regional dimension, which national level analysis does not properly capture in the EU. The gap and variation in regional competitiveness should stimulate a debate to what extent these gaps are harmful for their national competitiveness and to what extent the internal variation can be remediated.

Part of the explanation to the large inequalities within EU NUTS 2 regions may then have to do with the differences in competitiveness. An economic entity in region which has low competitiveness may not have similar opportunities as an economic entity in a highly competitive region. This fact remains and is confirmed. But what does it mean for efficiency in competitiveness? In the case of efficiency analysis of competitiveness and in time comparison analysis of change in 2013 to 2010, the results are just a little bit different. Why? The concept of competitiveness may then be important not only to evaluate why some regions grow faster than others, but also why some regions have a better and more efficient distribution of competitiveness over time than others. Is it a high level of competitiveness necessarily associated with a high level of efficiency, and vice versa? It may not always be the case because evaluated regions having lower level of inputs, resp. values of the sub-indices, these regions were able to achieve competitiveness at level of the RCI. The RCI value may not be high, and even in the less competitive regions is not, but it is necessary to compare the values of three inputs and one output. Very important is also the fact that with given level of inputs regions were able to achieve level of output, although less in 2013 than in 2010, overall it is possible to state that the regions with the production of output based on inputs operate more efficiently in 2013 than in the case of 2010, and otherwise. These results are not surprising. Finally, Tables 6 – 8 show reordered regions, from best to worst, the MPI score and the corresponding rank (low ranks are associated to high MPI scores, and also to high RCI score). Hereafter, these ranks are referred as 'reference ranks.' According to the efficiency analysis and derived results from the MPI solution, it emerges that 2013 to 2010 efficiency ratio of 212 NUTS 2 regions of the EU15 Member States range from 0.30506 – 212th position (UKL2 – East Wales) to 1.65698 – 1st position (GR24 – Sterea Ellada). 143 NUTS 2 regions of all 212 EU15 NUTS regions have recorded positive trend in competitiveness efficiency, 69 NUTS 2 regions of all 212 EU15 NUTS regions have achieved negative trend in competitiveness efficiency comparing 2013 to 2010 (see Tables 6 – 7). These results mean that 2/3 of all 212 EU15 NUTS regions make improvement in their competitiveness, i.e., in utilization of inputs for producing output.

Tables 6 – 8 show the MPI' scores for 2010-2013 and ranks for EU15 and EU12 NUTS 2 regions. In many countries, the capital region is far more competitive than the other regions in the same country and many countries have highly heterogeneous RCI scores. The gaps and variation in regional competitiveness should give rise to a debate on to what extent these gaps are harmful for their national competitiveness, and to what extent the internal variation can be remediated. The internal variation and heterogeneity also underlines the inevitable limits of a national level analysis. On the contrary, the RCI shows a more polycentric pattern with strong capital and metropolitan regions in many parts of Europe. Some capital regions are surrounded by similarly competitive regions, but in many countries the regions neighbouring the capital are far less competitive. A key question for the future is whether the strong performance of these capital and metropolitan regions will help to increase the performance of neighbouring regions through spillovers, or whether the gap between them and the other regions will grow.

According to the efficiency analysis and derived results from the MPI solution, it emerges that 2013 to 2010 efficiency ratio of 56 NUTS 2 regions of the EU12 Member States range from 0.72458 – 56th position (MT00 – Malta) to 6.35528 – 1st position (RO32 – București - Ilfov). 42 NUTS 2 regions of all 56 EU12 NUTS regions have recorded positive trend in competitiveness efficiency, 14 NUTS 2 regions of all 56 EU12 NUTS regions have achieved negative trend in competitiveness efficiency comparing 2013 to 2010 (see Table 8). These results mean that 2/3 of all 56 EU12 NUTS regions make improvement in their competitiveness, i.e., in utilization of inputs for producing output.

Based on the MPI results is clear, that the best efficiency changes in competitiveness comparing 2013 to 2010 were achieved by NUTS 2 regions belonging to the group of EU12 countries, i.e., new EU Member States than in the case of NUTS 2 regions belonging to the group of EU15 countries, i.e., old EU Member States. This fact is not surprising, because it has the key political implications and there are several reasons/factors for it, as follows:

1. New EU Member States constantly fall into the category of less developed and competitive states based on GDP per head in PPS – reason for inclusion of their NUTS 2 regions in the appropriate categorization stage of development.
2. Belonging of each region to relevant stage of development testifies to its competitive advantages and disadvantages and determines its weaknesses. Medium stage of development is associated with regional economies primarily driven by factors such as lower skilled labour and basic infrastructures. Aspects related to good governance and quality of public health are considered basic inputs in this framework. Intermediate stage of development is characterized by labour market efficiency, quality of higher education and market size, factors which contribute to a more sophisticated regional economies and greater potential for competitiveness. In the high stage of development, factors related to innovation, business sophistication and technological readiness are necessary inputs for innovation-driven regional economies (Annoni and Dijkstra, 2013).
3. The threshold defining the level of GDP in % of EU average has been taken as a reference as it is the criterion for identifying regions eligible for funding under the established criteria of the EU Regional Policy framework. European funds are an important tool for regional development and reducing economic, social and territorial disparities among European regions. Reducing disparities have a significant impact on competitiveness, and these two concepts are thus the EU complementary objectives. Of the total budget allocated to regional policy, substantial part goes just to NUTS 2 regions of EU12 countries whose development is thus supported.
4. New EU Member States are often significantly dependent on the exports into old EU Member States and on the flow of money for this exchange shift.

All these factors affect the convergence trend of new EU Member States and their regions to old EU Member States, and the growth in old EU Member States has implicative impact on growth in new EU Member States. This growth may have the same degree in EU12 countries as in EU15 countries, or rather is a higher and multiplied. Many of the differences in economic growth and quality of life within a country may be explained by the differences in competitiveness. Regions with more paved roads, with better institutions, with better business environment, and with better human capital, for example, may experience faster economic growth and a clearer reduction in poverty levels (Charles and Zegarra, 2014). All these trends and facts have very significant on competitiveness of all EU Member States and changes of its level and efficiency/inefficiency development. The gaps and variation in regional competitiveness should give rise to a debate on to what extent these gaps are harmful for their national competitiveness, and to what extent the internal variation can be remediated (Dijsktra, Annoni and Kozovska, 2011). The internal variation and heterogeneity also underlines the inevitable steps needed to make at national level to solve the main economic and social problems of their citizens. Just effective thematic policies and efficient use of public spending on the established aims will help to overall efficiency of the whole system, ensure desired outcomes.

Table 6. The MPI 2010-2013 Scores and Ranks for EU15 NUTS 2 Regions (I)

Rank	NUTS 2	OO MPI	Rank	NUTS 2	OO MPI	Rank	NUTS 2	OO MPI	Rank	NUTS 2	OO MPI
1	GR24	1,65698	28	SE11	1,13481	55	GR22	1,08050	82	DE73	1,05409
2	GR30	1,57816	29	SE23	1,13471	56	BE23	1,07965	83	DE22	1,05403
3	GR25	1,56626	30	FR62	1,13438	57	UKJ2	1,07877	84	DE14	1,05329
4	GR13	1,53967	31	SE33	1,13430	58	GR43	1,07674	85	BE35	1,05239
5	GR23	1,49430	32	AT34	1,13284	59	UKJ1	1,07577	86	BE10	1,05172
6	GR12	1,43491	33	ITF5	1,12879	60	AT13	1,07546	87	BE22	1,05056
7	GR11	1,43004	34	ITC3	1,12326	61	AT12	1,07546	88	ES43	1,05019
8	IE01	1,36030	35	SE32	1,12264	62	FR63	1,07465	89	DE24	1,04918
9	LU00	1,27070	36	AT33	1,12231	63	ITF1	1,07290	90	BE33	1,04906
10	ITF6	1,26236	37	SE21	1,12190	64	DED1	1,07288	91	DE94	1,04807
11	IE02	1,25403	38	PT18	1,11685	65	GR42	1,07174	92	UKK3	1,04784
12	ITF2	1,24338	39	UKJ4	1,11630	66	PT16	1,07127	93	UKD4	1,04750
13	ITF4	1,21203	40	ITE4	1,11570	67	BE25	1,07083	94	ES22	1,04680
14	ITF3	1,20136	41	PT15	1,11324	68	ITE1	1,06574	95	FR21	1,04612
15	AT21	1,18813	42	ES70	1,11268	69	ITE2	1,06574	96	FR72	1,04553
16	SE12	1,17649	43	AT32	1,10899	70	UKE3	1,06517	97	NL13	1,04528
17	PT17	1,17415	44	AT31	1,10356	71	BE32	1,06411	98	UKD2	1,04429
18	AT11	1,16904	45	ITD4	1,09299	72	FR53	1,06197	99	ITD3	1,04412
19	AT22	1,15519	46	FI13	1,08995	73	ITC1	1,06152	100	FR81	1,04233
20	NL23	1,15086	47	FI1A	1,08995	74	FI18	1,05925	101	NL42	1,04113
21	NL32	1,15086	48	FR61	1,08915	75	FI19	1,05847	102	BE34	1,03745
22	ITG1	1,15076	49	ITG2	1,08887	76	DED2	1,05764	103	DE60	1,03686
23	GR14	1,14893	50	BE21	1,08608	77	FR10	1,05638	104	NL21	1,03440
24	SE22	1,14500	51	ITE3	1,08304	78	FR23	1,05632	105	UKD3	1,03318
25	SE31	1,14101	52	DE71	1,08253	79	FR43	1,05539	106	DEA3	1,03238
26	DK01	1,14017	53	UKD5	1,08121	80	FR33	1,05539	107	ES13	1,03217
27	GR21	1,14014	54	NL34	1,08081	81	FR25	1,05529	108	FR71	1,03045

Source: Own calculation and elaboration, 2016.

Table 7. The MPI 2010-2013 Scores and Ranks for EU15 NUTS 2 Regions (II)

Rank	NUTS 2	OO MPI	Rank	NUTS 2	OO MPI	Rank	NUTS 2	OO MPI	Rank	NUTS 2	OO MPI
109	DE26	1.02696	136	FR24	1.01059	163	FR26	0.98961	190	UKH1	0.94559
110	ITD5	1.02603	137	FR82	1.01007	164	DEB3	0.98874	191	UKH2	0.94559
111	DE23	1.02598	138	FR52	1.00998	165	ES41	0.98736	192	UKH3	0.94559
112	ES12	1.02542	139	PT11	1.00828	166	FR41	0.98454	193	UKI	0.94559
113	DE80	1.02446	140	DK04	1.00692	167	NL41	0.98236	194	DEE0	0.94313
114	ITC4	1.02444	141	ITD1	1.00404	168	NL33	0.98123	195	ES42	0.94221
115	NL31	1.02383	142	FR42	1.00233	169	NL22	0.98016	196	FR30	0.93978
116	ITC2	1.02257	143	ES30	1.00050	170	DE91	0.97990	197	UKC2	0.93334
117	UKF3	1.02178	144	DE11	0.99882	171	UKL2	0.97952	198	UKL1	0.92240
118	ES23	1.02163	145	DE13	0.99816	172	DE30	0.97846	199	UKD1	0.91999
119	DE25	1.02024	146	DK03	0.99802	173	DE41	0.97846	197	ES51	0.91073
120	DED3	1.01938	147	DEA4	0.99654	174	DE42	0.97846	198	ES53	0.85069
121	UKF1	1.01861	148	DK05	0.99601	175	ES24	0.97827	199	ES62	0.82128
122	UKF2	1.01861	149	DE93	0.99572	176	UKG1	0.97676	200	PT30	0.79291
123	UKE4	1.01759	150	UKG2	0.99536	177	DEC0	0.97001	201	ITD2	0.79153
124	UKG3	1.01692	151	UKM2	0.99413	178	DE50	0.96980	202	FR92	0.75831
125	DEG0	1.01620	152	UKM3	0.99413	179	UKN0	0.96877	203	FI20	0.66462
126	DE21	1.01607	153	NL11	0.99328	180	DE72	0.96781	204	PT20	0.46734
127	DEA1	1.01602	154	DE27	0.99320	181	UKK1	0.96604	205	FR93	0.45303
128	FR22	1.01565	155	DEA2	0.99318	182	UKE2	0.96151	206	FR94	0.45303
129	DEA5	1.01502	156	GR41	0.99314	183	UKK4	0.95928	207	FR91	0.41708
130	DEB2	1.01473	157	DEF0	0.99306	184	UKE1	0.95906	208	ES63	0.32222
131	ES21	1.01445	158	UKJ3	0.99145	185	DE92	0.95872	209	ES64	0.32222
132	DEB1	1.01400	159	UKC1	0.99012	186	ES61	0.95797	210	DE91	0.31005
133	NL12	1.01354	160	ES11	0.99008	187	UKM5	0.95297	211	UKH1	0.30899
134	UKK2	1.01266	161	FR51	0.98991	188	ES52	0.95094	212	UKL2	0.30506
135	DE12	1.01172	162	DK02	0.98980	189	UKM6	0.94864			

Source: Own calculation and elaboration, 2016.

Table 8. The MPI 2010-2013 Scores and Ranks for EU12 NUTS 2 Regions

Rank	NUTS 2	OO MPI	Rank	NUTS 2	OO MPI
1	RO32	6.35528	29	PL62	1.11191
2	RO21	5.60765	30	PL41	1.11115
3	RO11	5.54264	31	CZ06	1.10565
4	RO42	5.50700	32	PL42	1.09618
5	RO31	5.21191	33	PL51	1.06126
6	RO41	5.14197	34	HU33	1.06126
7	RO22	5.08068	35	CZ03	1.05317
8	RO12	4.67104	36	CZ05	1.04763
9	BG33	3.82007	37	PL63	1.03866
10	BG41	3.61128	38	CZ02	1.03780
11	BG34	3.22278	39	PL61	1.02831
12	BG32	3.19500	40	CZ04	1.02593
13	BG42	2.82878	41	CZ01	1.02491
14	BG31	2.82755	42	PL33	1.00952
15	HU32	1.49790	43	CZ08	0.97023
16	SK03	1.24063	44	PL32	0.95733
17	SK04	1.19656	45	PL22	0.95543
18	PL43	1.17416	46	PL12	0.94667
19	SK01	1.14319	47	LT00	0.93848
20	PL11	1.14262	48	PL34	0.89365
21	PL21	1.14171	49	EE00	0.88625
22	SI01	1.14060	50	CY00	0.88379
23	SI02	1.13250	51	LV00	0.88004
24	HU23	1.13051	52	HU21	0.87590
25	SK02	1.12529	53	PL31	0.85575
26	PL52	1.12154	54	HU22	0.85309
27	CZ07	1.11993	55	HU10	0.81262
28	HU31	1.11376	56	MT00	0.72458

Source: Own calculation and elaboration, 2016.

With respect to gained results, it is possible to create four groups of European regions – classification of each region to relevant group is based on two dimensions of the MI – Catch-up effect and Frontier-shift effect. Classification of NUTS 2 regions separately for EU15 and EU12 countries with respect to the nature of technical and technological change is illustrated as Scatterplots in Figure 2 and Figure 3. Most of evaluated regions within both groups of countries belong to the first group, least of evaluated regions belong to the third group. From classification is also evident how many regions within EU15 and EU12 countries belong to each group – it is possible to state that results in over-time efficiency deterioration and improvement are more or less balanced across EU15 and EU12 countries comparing 2013 to 2010.

On Scatterplots, location of all European regions is recorded with respect to results, resp. their values of technical efficiency change (Catch-up effect) and technological efficiency change (Frontier-shift effect). Evaluated NURS 2 regions are divided in two groups – EU15 Member States and EU12 Member States for better comparison of common features and differences. It is convenient to remind that Catch-up and Frontier-shift in values of 1.000 mean no productivity change, values lower than 1.000 mean productivity deterioration and values higher than 1.000 mean productivity improving.

From this point of view, it is possible to divide European regions in four categories, resp. quadrant. In period 2010–2013, most of European regions were moving in I. quadrant what means that value of technical efficiency change (EFCH), i.e., Catch-up is higher than 1.000 and value of technological efficiency change, i.e., Frontier-shift (FS) is lower than 1.000. This fact is not surprising, but expected, because it means that efficiency change of competitiveness comparing 2013 to 2010 is (in the case of these regions) caused by change in relative efficiency of regions to other regions in sample group. Only a small contribution to the overall change was caused by technological efficiency. It means using of small progress in technology, innovation and improvement, i.e., with respect to the theory of growth – increase in steady state or effective options frontier of region. The exact values for EFCH and FS of NUTS 2 regions belonging to I. quadrant (separately for EU12 and EU15 countries) are seen in Table 1 in Annex. In the case of EU12 countries, other group with a high proportion of regions is II. quadrant, but number of NUTS 2 regions belonging to this quadrant is not so high as in the case of I. quadrant. In the case of EU15 countries, number of NUTS 2 regions belonging to I. quadrant is very similar as in the case of EU12 countries. II. quadrant is characterized by high level of Frontier-shift effect and high level of Catch-up effect, i.e., value of technical efficiency change is higher than 1.000 and value of technological efficiency is also higher than 1.000. For other III. quadrant is typical high level (higher than 1.000) of Frontier-shift effect and low level (lower than 1.000) of Catch-up effect. In the field of understanding of technical and technological efficiency change and their importance for the nature of overall change, interpretation of III. quadrant is the exact opposite than interpretation of I. quadrant. It is thus not surprising, that smaller number of regions are classified in III. quadrant than in I. quadrant – with respect to possibilities of NUTS 2 regions. Number of regions in III. quadrant is greater for EU15 countries than for EU12 countries, what is expected with respect to their stage of development. Last but not least, IV. quadrant is specified by low level of both dimensions of the MI. Value of technical efficiency change (EFCH), i.e., Catch-up is lower than 1.000 and value of technological efficiency change, i.e., Frontier-shift (FS) is also lower than 1.000. No NUTS 2 region of EU12 countries belong to IV. quadrant, but on the contrary, quite a large number of NUTS 2 regions of EU15 countries are ranked within this IV. quadrant. It could be due to the fact that EU15 countries have more regions, as well as high disparities and not only among regions, but especially within the states.

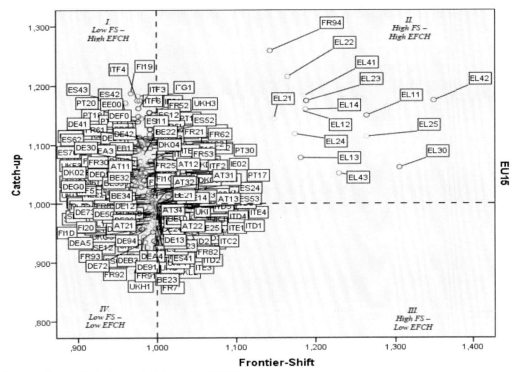

Source: Own calculation and elaboration, 2016.

Figure 2. Scatterplot for EU15 NUTS 2 Regions.

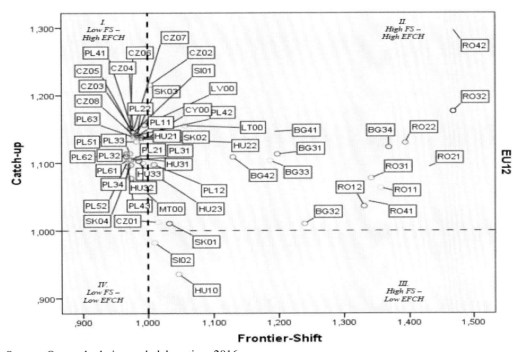

Source: Own calculation and elaboration, 2016.

Figure 3. Scatterplot for EU12 NUTS 2 Regions.

Via illustration of Figure 2 and Figure 3, information about differences in efficiency of competitiveness recorded by the MPI 2010-2013 between group of EU12 and EU15 countries are confirmed. Once again it should be emphasized, that most of NUTS 2 regions of EU12 countries are located in quadrants with low level of FS, and high or level of EFCH. It means that efficiency change is change in the relative efficiency of evaluated region in relation to other regions, due to the production possibility frontier in period 2010-2013, i.e., technical efficiency change. This fact is not so positive information because it means that regions extract their efficiency based on shifts in sources of competitiveness, i.e., they make changes in composition and quantity of sources based on exchange business with other countries. Character of technical efficiency change thus contribute only to quantitative based economic growth which has its limits; this is disconcerting with reference to limited sources, utilization of sources and possibility/impossibility to their recovery. Results for EU15 countries, resp. their NUTS 2 regions are more or less similar, but it is necessary to note that also significant number of EU15 regions is located in quadrants with high level of FS, and high level of EFCH. It means that significant part efficiency change is caused especially by the change in the production possibility frontier as a result of the technology development in period 2010-2013, i.e., technology frontier shift. This fact is positive information with respect to factors of competitiveness, it signifies that regions are able to utilize their internal factor endowment in effective way and are able to apply technological progress for boosting of their competitive advantages, i.e., they contribute thus to qualitative based economic growth and it is option how to raise the steady state. What are factors (internal or external) having impact on steady state? Traditional factors such as physical infrastructure and access to land, labour, materials, markets and capital remain the basic determinants of competitiveness. However, the economy has changed, and so has regional policy. In the days when smoke-stack industries sat protected by national tariff barriers, regional policy was mainly about hard infrastructure – new factories and roads bestowed by governments, gifts from outside the region itself. Today the response involves upgrading the business environment through "soft infrastructure." Less tangible assets need to be cultivated, that enhance territorial capital and enable a region to realise its own potential (ESPON, 2006b). The exact formula for efficiency development in competitiveness will depend on the particular region. For example, in less prosperous states and regions gaps in health care can be a barrier to economic development. Promotion of social inclusion and sustainable communities may be particularly important in metropolitan regions. In addition, the urban environmental quality has become a factor of more importance. Some of these key "modern" drivers of competitive performance are now discussed.

Regional Competitiveness Index and Human Capital Dimensions

With respect to dimensions of human capital and their impact on competitiveness, following part describe consequences of the RCI pillars connecting with human capital, i.e., Health, Quality of Primary and Secondary Education and Higher Education/Training and Lifelong Learning. Why these pillars are important for regional competitiveness and what impact have on the RCI results in the case of the EU NUTS 2 regions? It is a subject of next synthetic part of the chapter.

Health

This pillar is devoted to the description of human capital in terms of health condition and well-being, with special focus on the workforce. The 2006 Community Strategic Guidelines on Cohesion (Official Journal of the European Union, 2006) underline that a healthy workforce is a key factor in increasing labor market participation and productivity and enhancing competitiveness at national and regional level. They point out to major differences in health status and access to health care across the EU regions. Good health conditions of the population lead to greater participation in the labor force, longer working life, higher productivity and lower healthcare and social costs. Indicators to measure some of these aspects, available from Eurostat at the NUTS 2 regional level, are follows: 'hospital beds,' 'infant mortality,' 'cancer death,' and 'heart disease death rates' are meant to describe the outcomes of the health system. 'Road fatalities' and 'suicide rates' refer to aspects of the population's well-being from a more social point of view. The intention is to measure the health system's effectiveness with regard to the population's health and well-being rather than to the level of inputs. Many aspects – not strictly dependent directly upon the health system such as food and/or smoking habits, population density, quality and safety of roads, etc. – may affect the indicators included in the pillar. The pillar aims to describe a wider concept: the competitiveness of the workforce in terms of healthy living conditions (Annoni and Kozovska, 2010).

Southern European and Scandinavian regions have very low numbers of hospital beds. Road fatalities present biggest problem in Southern European regions (Spanish, Greek and Portuguese) as well as in the Baltic countries. UK regions are among the ones with the lowest number of road fatalities. Most of the Scandinavian and Greek regions have very high healthy life expectancy while regions in the Baltic States, Finland, Hungary and Slovakia are among the ones with the lowest performance. Infant mortality is highest in Eastern European regions, Bulgaria and Romania specifically, while best performers are regions in Italy, Greece, Germany and United Kingdom. Cancer rate is highest in a number of Eastern European regions (Romanian, Hungarian, Bulgarian, Baltic) while best performers are parts of Italy, Sweden and Finland. Similarly, heart diseases are most common in Eastern Europe while most rare in Spanish, Portuguese and Southern French regions. Suicide rates are very low in Southern European regions and very high in Northern European regions (Annoni and Kozovska, 2010).

Quality of Primary and Secondary Education

High levels of basic skills and competences increase the ability of individuals to subsequently perform well in their work and to continue to tertiary education. A number of studies have found a significant positive association between quantitative measures of schooling and economic growth (Sianesi and Reenen, 2003; Krueger and Lindahl, 2001; Hanushek and Wößmann, 2007). In addition, a recent paper by the Lisbon Council, argues that human capital is best managed at the regional level, underlining that relevance of the regional dimension (Ederer et al., 2010). To capture this dimension, the RCI is, in this part, focused on compulsory education outcomes as an indication of effectiveness and quality of the educational system across the EU Member States. To this aim, it has taken into account the performance of students in the OECD Programme for International Student Assessment (PISA) wave. Recent OECD studies (OECD, 2010 a, b) relate cognitive skills measured by PISA to economic growth and finds that relatively small improvements in the skills of a nation's labour force can have substantial gains in terms of current GDP and future well-being. PISA indicators make it possible to identify the share of pupils, 15-year-old, who have a low level of basic skills in reading, math and science. Pupils who fail to reach higher levels can be considered to be inadequately prepared for the challenges of the knowledge society and for lifelong learning, thus indicating a lower potential in terms of human capital. In order to describe educational input factors, it also consider indicators related to teacher to pupil ration, public expenditure on compulsory education and financial aid available for students. Investment in education can be considered as an essential element in guaranteeing good quality of the educational system. Participation in early childhood education has become one of the new EU benchmarks in the field of education and training. Several studies have pointed out to the positive effects of early childhood education from an educational and social perspective as it can counter potential educational disadvantages of children, coming from unfavorable family situations (NESSE, 2009; European Commission, 2009). In the RCI are thus included potential indicators measuring this aspect, i.e., 'low achievers in reading,' 'low achievers in math,' 'low achievers in science.'

Bulgaria and Romania are the countries with the highest percentage of low achievers in reading, math and science. Finland is the top performer in all three fields, together with Ireland (for reading), the Netherlands (for math), and Estonia (for science). (Annoni and Kozovska, 2010).

Higher Education/Training and Lifelong Learning

The contribution of education to productivity and economic growth has been widely researched in the last decades. Knowledge-driven economies based on innovation require well-educated human capital, capable to adapt, and education systems which successfully transmit key skills and competences. A clear picture of the economic benefits of education can be found in the most current release of the OECD publication Education at a Glance 2009 (OECD, 2009). As also underlined by the Lisbon Council president (Hofheinz, 2009), the main findings of the OECD report are straightforward: investment in educations pays always, for the individual and for society at large. Further, a stream of research literature in the past two decades has shown that the quality of human resources is not only directly involved in knowledge generation but plays a crucial role for applying and imitatatin technologies developed somewhere else (Azariadis and Drazen, 1990). It is clear that this pillar plays a key role in describing competitiveness.

Variables traditionally used for measuring educational quality are levels of educational attainment of the population, number of years of schooling of the labour force or literacy rates (Psacharopoulous, 1984). Participation in education throughout one's life has also been deemed essential for the continuous upgrade of the skills and competences of workers in order to assist them in handling the challenges of continuously evolving technologies. In this pillar, these aspects are captured by proposing to include indicators on levels of tertiary educational attainment, participation in lifelong learning among the population as well as percentage of young people who have left the educational system at an earlier stage.

Furthermore, an indicator of geographical accessibility to higher education institutions is proposed as a relevant factor, especially at the regional level. All these indicators are available at the required NUTS 2 regional level. The analysis has been complemented by adding a fifth indicator to take into account the expenditure on tertiary education (Annoni and Kozovska, 2010).

With regards to tertiary education attainment, from the RCI results it is evident that a number of UK regions are performing very well while the northern regions of Romania show some of the lowest performance. Northern European regions perform best on the lifelong learning indicator while we can see parts of Romania, Bulgaria and Greece having the lowest percentage of the population participating in lifelong learning activities. A number of Polish regions perform very well on the indicator on early school leavers while Mediterranean regions, especially in Portugal and Spain, are lagging significantly behind. German regions demonstrate a very dense network of universities while Greek regions have the worst accessibility to universities. Denmark and Finland are the countries with highest expenditure on tertiary education as percentage of GDP while Bulgaria and Italy have the lowest. In general, we could see a rather distinct division between the performance of Northern and Southern European regions in terms of the quality of higher education and training systems (Annoni and Kozovska, 2010).

CONCLUSION AND RECOMMENDATIONS

Globalisation makes international competitiveness a key concern in regional development. A European dimension is becoming essential for effective development of smaller or larger NUTS 2 regions. Large geographic, demographic and cultural diversity of the EU brings also differences in socio-economic position of the EU Member States, and especially their regions. Different results in economic performance and living standards of the population indicate the status of the competitiveness of every country and its regions. Each territory should know were lying its competitive advantages and advantages and try to strengthen its advantages and reduce its disadvantages, i.e., key factors of competitiveness. Understanding of the position of each region in comparison to others can thus highlight new potentials for development. Key options have to be explored from a European perspective; understanding the larger territorial context makes it easier to spot new opportunities and under-used potentials (ESPON, 2006b). Awareness raising, dialogue and involvement are vital parts of the process of empowering policy makers and practitioners at different levels, so that they can exploit comparative advantages and add value through targeted territorial cooperation with other regions. By mobilising the existing potential for growth in all regions, cohesion policy can both improve the geographical balance of economic development, and also increase growth in Europe as a whole and support thus the level of competitiveness competitiveness and improve the level of efficiency in public spending. Mutual relationship between two significant themes, i.e., efficiency and competitiveness in the context of national economies was motivation for this chapter. Timeliness and importance of the topic for regional policy-makers is high; see e.g., (Camanho and Dyson, 2006; Hančlová, 2013; Khan and Soverall, 2007; Melecký, 2013a, b; Mihaiu, Opreana and Cristescu, 2010; Staníčková,

2013). All regions have their part to play, especially those where the potential for higher productivity and employment is greatest.

Comparisons can enable NUTS 2 regions to identify their strengths and weaknesses in a European context and to enrich their development strategies, project ideas and cooperation arrangements. In the core of Europe all types of regions are doing well with regard to both restructuring potential and economic situation, indicating a high potential regional competitiveness. In particular, countries in the core of Europe, Ireland, large parts of the UK, the regions in Western France, Spain, and the capital regions of Norway, Sweden and Finland are well placed in the field of competitiveness. Economically weaker regions, with deficits concerning their restructuring potential, can be found in most countries, in particular in Central and South-Eastern areas of Germany, Poland, Czech Republic, Austria, Italy, Hungary, Romania, Bulgaria, Greece and Cyprus. Many of these areas also have a labour market classification below the European average. Cyprus, however, differs in this regard from its neighbouring countries. The European territorial pattern seems mainly shaped by different national levels. In addition, a substantial difference exists between rural and urban areas. The more urbanised regions have as expected the best potentials for pursuing strategies of innovation and knowledge-economy based on a particularly creative, segment of the competitive society. Some areas more than others may take on the idea of new/creative industries on the one side, or on the other side on the idea of traditional industries, as a motor for economic development and innovation. As development potentials and opportunities and the interplay of development trends and policies differ between areas, there is no one-size-fits-all solution. Each NUTS 2 region must make its own decisions about the right combination of policy objectives in the field of competitiveness that will guide its development.

Bringing together different development factors which illustrate single aspects of competitiveness gives a first impression of the overall international competitiveness of European regions and shows the diversity that exists within the EU territory (ESPON, 2006a). Among the important driving forces influencing future territorial development are demographic development (including migration), economic integration, transport, energy, agriculture and rural development, climate change, further EU enlargements and territorial governance. Very important role plays exogenous factors having impact on regional competitiveness, as mentioned (ESPON, 2006b). Current theories of regional competitiveness emphasise the significance of "soft" factors such as human, cultural (knowledge and creativity) and socio-institutional capital, environmental quality, etc. Wide range of soft location factors are thus of increasing importance. "Soft" factors like human capital dimensions, governance, culture and natural environment are part of territorial potentials and offer synergies for the jobs and growth agenda. Knowledge and other human capital elements, as intellectual capital components play crucial role in establishing and maintenance of competitive advantage of territories. And, while attributes pertaining to products and services in the market may for a while provide for competitive advantages, knowledge as an element of human capital and other human and intellectual capital components are source of key competences and a factor that has crucial influence on creating a long-term sustainable competitive advantages. The potentials for these "soft factors" differ widely between areas. Quality living environments, and access to environmental and cultural amenities are among factors that attract investment and people to a location what is very important for competitiveness for each NUTS 2 regions and its competitive advantage and factor endowment. Currently hazards do not undermine the competitiveness of a region. Only a few places have very low exposure to the main natural and technological hazards in Europe, and climate change is expected to increase the risk of hazards in the future. To gaze into the future, it is necessary to understand the driving forces that shape territorial development and various possible future developments and interrelations with the territory each driving force might bring. Bringing them together into integrated prospective scenarios is then the final challenge.

REFERENCES

Annoni, A and Kozovska, K. (2010). *EU Regional Competitiveness Index 2010*. Luxembourg: Publication Office of the European Union.

Annoni, P. and Dijkstra, L. (2013). *EU Regional Competitiveness Index 2013*. Luxembourg: Publication Office of the European Union.

Azariadis, C. and Drazen, A. (1990). Threshold externalities in economic development. *Quarterly Journal of Economics*, 105 (2): 501-526.

Camanho, A.S. and Dyson, R.G. (2006). Data envelopment analysis and Malmquist indices for measuring group performance. *Journal of Productivity Analysis,* 26 (1): 35-49.

Caves, D. W., Christensen, L. R. and Diewert, W. E. (1982). The Economic Theory of Index Numbers and the Measurement of Input, Output, and Productivity. *Econometrica*, 50 (6): 1393-1414.

Charles, V. and Zegarra, L.F. (2014). Measuring regional competitiveness through Data Envelopment Analysis: A Peruvian case. *Expert Systems with Applications*, 41: 5371-5381.

Charnes, A., Cooper, W.W. and Rhodes, E. (1978). Measuring the efficiency of decision making units. *European Journal of Operation Research*, 2 (6): 429-444.

Cooper, W.W., Seiford, L.M. and Tone, K. (2007). *Introduction to Data Envelopment Analysis and its Uses with DEA-solver Software and References*. New York: Springer.

De la Fuente, A. and Ciccone, A. (2002). *Human capital in a global and a global based economy* [online]. Available at: http://www.dbresearch.com/PROD/DBR_INTERNET_EN-PROD/PROD0000000000190080.PDF [18.08. 2009].

Deutche Bank Research (2005). *Human capital is the key to growth* [online]. Available at: http://www.dbresearch.com/PROD/DBR_INTERNET_ENPROD/PROD0000000000190080.PDF.

Dijkstra, L., Annoni, A. and Kozovska, K. (2011). A New Regional Competitiveness Index: Theory, Methods and Findings. European Commission, Regional Policy *Working Paper n° 2/2011*.

Djurica, M., Djurica, N. and Janicic, R. (2014). Building Competitive Advantage Through Human Capital. In: *The Clute Institute International Academic Conference*. Munich: The Clute Institute, pp. 553-558.

Ederer, P., Schuller, P. and Willms, S. (2010). *Human Capital Leading Indicators: How Europe's Regions and Cities Can Drive Growth and Foster Social Inclusion*. Lisbon: Lisbon Council report.

European Commission (2009). *Progress towards the Lisbon Objectives in Education and Training: Indicators and benchmarks 2009*. Brussels: Commission staff Working Document.

European Spatial Planning Observation Network (2006a). Mapping regional competitiveness and cohesion European and global outlook on territorial diversities. *ESPON Briefing 2*.

European Spatial Planning Observation Network (2006b). Territory matters for competitiveness and cohesion. Facets of regional diversity and potentials in Europe. *ESPON Synthesis Report III*.

Färe, R., Grosskopf, S. and Lovell, C. (1994a). *Production Frontiers*. Cambridge: Cambridge University Press.

Färe, R., Grosskopf, S., Norris, M. and Zhang, Z. (1994b). Productivity Growth, Technical Progress and Efficiency Change in Industrialized Countries. *The American Economic Review*, 84 (1): 66-83.

Fojtíková, L., et al. (2014). *Postavení Evropské unie v podmínkách globalizované světové ekonomiky*. Ostrava: VŠB-TU Ostrava. [*The EU's Position in Terms of the Globalized World Economy*. Ostrava: VŠB-TU Ostrava].

Hančlová, J. (2013). Econometric Analysis of Macroeconomic Efficiency Development in the EU15 and EU12 Countries. In: *Proceedings of XI. International Scientific Conference Liberec Economic Forum 2013*. Liberec: Technical University of Liberec, pp. 166-175.

Hanushek, E. A. and Wößmann, L. (2007). *Education Quality and Economic Growth*. Washington DC: World Bank report.

Hofheinz, P. (2009). *EU 2020: Why skills are key for Europe's future* [online]. Available at: http://www.lisboncouncil.net/publication/ publication/54-skillseuropesfuture.html.

Institute for Management and Development (2012). *IMD World Competitiveness Yearbook 2012*. Lausanne: IMD.

Khan, J. and Soverall, W. (2007). *Gaining Productivity*. Kingston: Arawak Publications.

Krueger, A. and Lindahl, M. (2001). Education for growth: Why and for whom? *Journal of Economic Literature*, 39 (4): 1101-1136.

Krugman, P. (1994). Competitiveness: A Dangerous Obsession. *Foreign Affairs*, 73 (2): 28-44.

Mandl, U., Dierx, A. and Ilzkovitz, F. (2008). *The effectiveness and efficiency of public spending*. Brussels: European Commission-Directorate General for Economic and Financial Affairs.

Matovac, A., Bilas, V. and Franc, S. (2010). Understanding the Importance of Human Capital and Labour Market Competitiveness in the EU Candidate Countries and Selected EU Members. *Ekonomska misao i praksa*, 19 (2): 359-382.

Melecký, L. (2013a). Comparing of EU15 and EU12 Countries Efficiency by Application of DEA Approach. In: Vojáčková H (ed.). *International Conference Mathematical Methods in Economics 2013*. Jihlava: College of Polytechnics, pp. 618-623.

Melecký, L. (2013b). Use of DEA Approach to Measuring Efficiency Trend in Old EU Member States. In: *Proceedings of XI. International Scientific Conference Liberec Economic Forum 2013*. Liberec: Technical University of Liberec, pp. 381-390.

Melnikas, B. (2011). Knowledge economy: Synergy effects, interinstitutional interaction and internationalization processes. *Inzinerine Ekonomika-Engineering Economics*, 22 (4): 367-379.

Mihaiu, D.M., Opreana, A. and Cristescu, M.P. (2010). Efficiency, effectiveness and performance of the public sector. *Romanian Journal of Economic Forecasting*, 1 (4): 132-147.

Molle, W. (2007). *European Cohesion Policy*. London: Routledge.

NESSE (2009). Early childhood education and care [online]. Available at: http://www.nesse.fr/nesse/activities/reports/ececreport-pdf.

OECD (2009). *Education at a Glance 2009*. Paris: OECD.

OECD (2010a). *Education at a Glance 2010*. Paris: OECD.

OECD (2010b). *The High Cost of Low Educational Performance*. Paris: OECD.

Official Journal of the European Union (2006). *Council Decision of 6 October 2006 on Community strategic guidelines on cohesion (2006/702/EC)* [online]. Available at: http://ec.europa.eu/regional_policy/sources/docoffic/2007/osc/l_29120061021en00110032.pdf.

Porter, M. E. (2003). The Economic Performance of Regions. *Regional Studies*, 37 (6/7): 549-578.

Porter, M.E. (1990). *The competitive advantage of nations*. New York: The Free Press.

Psacharopoulous, G. (1984). The contribution of education to economic growth: international comparisons. In: Kendrick, J. (Ed.) *International comparisons of productivity and causes of slowdown*. Ballinger: American Enterprise Institute.

Sianesi, B. and J. V. Reenen (2003). The Returns to Education: Macroeconomics. *Journal of Economic Surveys*, 17 (2): 157-200.

Staníčková, M. (2013). Measuring of efficiency of EU27 countries by Malmquist index. In: *Proceedings of 31th International Conference Mathematical Methods in Economics 2013*. Jihlava: College of Polytechnics, pp. 844-849.

Staníčková, M. and Skokan, K. (2013). Multidimensional Approach to Assessment of Performance in Selected EU Member States. *International Journal of Mathematical Models and Methods in Applied* Sciences, 1 (7): 1-13.

Toloo, M. (2012). Alternative solutions for classifying inputs and outputs in data envelopment analysis. *Computers and Mathematics with Applications*, 63: 1104-1110.

World Economic Forum (2015). *The Global Competitiveness Report 2015 – 2016*. Lausanne: WEF.

Chapter 13

THE COX-INGERSOLL-ROSS INTEREST RATE MODEL REVISITED: SOME MOTIVATIONS AND APPLICATIONS

R.-V. Arévalo*, *J.-C. Cortés*† *and R.-J. Villanueva*‡
Instituto Universitario de Matemática Multidisciplinar,
Universitat Politècnica de València, Valencia, Spain

Abstract

We revisit the so-called Cox-Ingersoll-Ross (CIR) interest rate model. We pay particular attention to two different ways of motivating this stochastic model starting from its deterministic model counterpart. Afterwards, we explain two powerful techniques to estimate model parameters. Finally, the results are applied to model the 1-month Euribor interest rate from a real sample. Some measures of goodness-of-fit of the mathematical model are included.

Keywords: Euribor interest rate, Geometric Brownian Motion, Cox-Ingersoll-Ross model, Itô-type stochastic differential equations, maximum likelihood method

1. Introduction

The study of interest rates is a very significant issue whose modelling is not trivial since its behaviour depends on numerous key economic factors. Interest rates have a major influence on the economy. Some important examples include changes in the investment companies, fluctuations on the stock prices, level of savings in both companies and families, just to mention a few.

One of the most important effects can be observed when the interest rate remains high for a long time. This causes the reduction of companies' profits resulting in a decrease in the stock market as well. However, this also leads to the rise in prices for both loans

---

*E-mail address: ruben1.6180@gmail.com
†E-mail address: jccortes@imm.upv.es
‡E-mail address: rjvillan@imm.upv.es

and credits and therefore, a reduction in the consumption by companies and families. To the contrary, as the interest rate decreases for an extended period of time it encourages domestic consumption and, boosts economy.

The aforementioned points necessitate reasons to motivate the search for mathematical models which are able to predict interest rate dynamics. In this contribution, we revisit the so-called Cox-Ingersoll-Ross (CIR) interest rate model [1] to show how it can be applied to describe the dynamics of Euribor interest rate. CIR model is based on Itô stochastic differential equations (s.d.e.'s) [2].

The main contribution of this chapter is the development of a tailor-made parameter estimation method to the Cox-Ingersoll-Ross model. This method can be considered as an adaptation of the technique shown in [3] for another short-term interest rate model, usually referred to as Vasicek method [4].

This chapter is organized as follows: Section 2 is addressed to motivate the CIR model from its deterministic counterpart. Section 3 is addressed to show how the main statistical properties, such as mean and variance functions of the solution to the CIR model can be obtained even though an explicit solution of the CIR model is not available. The mean and variance functions can be obtained by applying the Itô Lemma. In Section 4, two statistical methods are developed to provide model parameter estimation. The first method is the maximum likelihood estimation method and the second one is a non-parametric technique. In Section 5 we validate the model using some appropriate statistical measures. Section 6 illustrates how the theoretical results introduced in previous sections can be applied to model and predict the 1-month Euribor interest rate. This application is validated using some goodness-of-fit measures. Conclusions are drawn in Section 7.

2. Motivating the CIR Stochastic Model

Throughout this section we introduce two different approaches in order to motivate the CIR model. As we shall see later, the first approach is based on considering perturbations in a classical deterministic interest rate model. The second approach leads to CIR interest rate model as the limit of a discrete stochastic model with adequate transition probabilities.

Numerous models have been proposed to describe interest rate dynamics. The majority of them are referred to as mean-reverting models, i.e. they tend to stabilize in the long-term. The following deterministic interest rate model can be considered as the cornerstone of mean-reverting models

$$\begin{cases} \mathrm{d}r(t) &= \alpha(r_e - r(t))\,\mathrm{d}t, \quad \alpha > 0, \\ r(0) &= r_0, \end{cases} \tag{1}$$

where $r(t)$ denotes the interest rate at the time instant $t > 0$, r_0 is the initial interest rate and α can be interpreted as the mean-reversion velocity of adjustment of $r(t)$ to the long-term interest rate r_e. This latter assertion is based on the fact that

$$r(t) = r_e + (r_0 - r_e)e^{-\alpha(t-t_0)} \xrightarrow[t \to \infty]{} r_e. \tag{2}$$

Since the long-term interest rate r_e is not known in a deterministic way, in the seminal contribution [4] the following stochastic model was proposed

$$\begin{cases} \mathrm{d}r(t) &= \alpha(r_e - r(t))\,\mathrm{d}t + \sigma \mathrm{d}B(t), \quad \alpha, \sigma > 0, \\ r(0) &= r_0, \end{cases} \quad (3)$$

where $B(t)$ denotes the brownian motion also termed the standard Wiener stochastic process. This formulation can be justified introducing the following perturbation $r_e \to r_e + \sigma B'(t)$ in the deterministic model (3). Here, σ parameter is referred to as the local volatility and $B'(t)$ denotes the white noise stochastic process [2].

Since volatility hardly ever remains constant another alternative to model (3) has been proposed. In this regard, the Cox-Ingersoll-Ross model is likely the most popular in applications. In this model it is assumed that σ depends on t, namely $\sigma(t) = \sigma\sqrt{r(t)}$. This hypothesis is based on the fact that brownian motion $B(t)$ has as standard deviation $\sqrt{t}$ on the time interval $[0, t]$, $t > 0$. Thus, it makes sense that the volatility behaves like the brownian motion. This yields the so-called Cox-Ingersoll-Ross model [1]

$$\begin{cases} \mathrm{d}r(t) &= \alpha(r_e - r(t))\,\mathrm{d}t + \sigma\sqrt{r(t)}\mathrm{d}B(t), \quad \alpha, \sigma > 0, \\ r(0) &= r_0, \end{cases} \quad (4)$$

Many generalizations of the CIR model have been proposed since the seminal contribution by [1]. In [5] a comparative study of 8 short-term interest rate models can be found. Here, we point out the CEV model where it is assumed that $\sigma(t) = \sigma r(t)^\lambda$, $\lambda > 0$. This generalization offers more flexibility when fitting real interest rate samples. In [6], a comparative study to several interest rate models is shown. The study is done by comparing implied parametric density with the same density estimated nonparametrically. A comprehensive study of short-term interest rate models can be found in [7, 8].

In [2] one can find a good motivation for the CIR interest rate model. As we shall show below, it is based on interest rate's changes for small time intervals Δt. Let us suppose that $r(t)$ represents the instantaneous interest rate for the time $t + \Delta t$, and let us denote by $\Delta r = r(t + \Delta t) - r(t)$. Then, there are three possibilities regarding these changes or increments of interest rate: increase, remain constant or decrease. These changes consist of $(\Delta r)_1 = 1$, $(\Delta r)_2 = 0$, and $(\Delta r)_3 = -1$, respectively. The next step is to find out their probabilities which thereafter will be denoted by p_1, p_2 and p_3, respectively. For convenience, let us assume that

$$\begin{cases} (\Delta r)_1 = 1 & \Rightarrow p_1 = \left(\frac{\sigma^2 r}{2} - \frac{\alpha(r_e - r)}{2}\right)\Delta t, \\ (\Delta r)_2 = 0 & \Rightarrow p_2 = 1 - p_1 - p_3, \\ (\Delta r)_3 = -1 & \Rightarrow p_3 = \left(\frac{\sigma^2 r}{2} + \frac{\alpha(r_e - r)}{2}\right)\Delta t. \end{cases} \quad (5)$$

The assignment of these probabilities has an intuitive interpretation, namely, p_1 and p_3 are centred about the average value of square of the diffusion coefficient, $\sigma\sqrt{r(t)}$, being its displacement the average value of the deterministic coefficient, $\alpha(r_e - r(t))$. Naturally, the value of p_2 is determined taking into account that $p_1 + p_2 + p_3 = 1$.

Therefore, the first and second statistical moments of increments Δr are directly computed from probabilities p_i, $1 \leq i \leq 3$,

$$\mathbb{E}[\Delta r] = \alpha(r_e - r)\Delta t, \quad \text{and} \quad \mathbb{E}[(\Delta r)^2] = \sigma^2 r \Delta t. \quad (6)$$

This is consistent with the general results exhibited in [2, p.140] where it is proven that $\mathbb{E}[(\Delta r)/(\Delta t)]$ corresponds to the coefficient of dt and $\sqrt{\mathbb{E}[(\Delta r)^2/\Delta t]}$ corresponds to the coefficient of $dB t$.

3. Solving the Stochastic Interest Rate Model

As it happens in dealing with deterministic differential equations, closed form solution stochastic process of many s.d.e.'s are not available. However, sometimes it is possible to calculate the mean and variance functions of the solution. Fortunately, this is the case for the CIR model. As we shall show below, this can be done taking advantage of Itô lemma and the crucial properties of Itô integral.

In order to the compute the expectation, $\mathbb{E}[r(t)]$, of the solution stochastic process to model (4), let us write it in the following equivalent integral form

$$\begin{aligned} r(t) &= r_0 + \alpha \int_0^t r_e \, ds - \alpha \int_0^t r(s) \, ds + \sigma \int_{r_0}^r \sqrt{r(s)} dB(s) \\ &= r_0 + \alpha r_e t - \alpha \int_0^t r(s) \, ds + \sigma \int_{r_0}^r \sqrt{r(s)} dB(s). \end{aligned} \quad (7)$$

We take the expectation operator in this latter expression and use the following properties of Itô integrals

$$\mathbb{E}\left[\int_a^b g(t) dt\right] = \int_a^b \mathbb{E}[g(t)] \, dt, \quad \mathbb{E}\left[\int_a^b g(t) dB(t)\right] = 0, \quad (8)$$

this leads

$$\mathbb{E}[r(t)] = r_0 + \alpha r_e t - \alpha \int_0^t \mathbb{E}[r(s)] \, ds. \quad (9)$$

Differentiating this expression with respect to t, the following initial value problem is obtained

$$\begin{cases} \dfrac{d\left(\mathbb{E}[r(t)]\right)}{dt} &= \alpha r_e - \alpha \mathbb{E}[r(t)], \\ r(0) &= r_0. \end{cases} \quad (10)$$

The solution of this deterministic initial value problem is well-known

$$m_1(t) = \mathbb{E}[r(t)] = r_e + (r_0 - r_e) e^{-\alpha t} \xrightarrow[t \to \infty]{} r_e. \quad (11)$$

Here, it is interesting to observe that the *averaged* value of the solution stochastic model (4) matches the solution of the deterministic model (1) and, both share the same long-term behaviour (see (2)).

Since

$$\mathbb{V}[r(t)] = \mathbb{E}\left[(r(t))^2\right] - (\mathbb{E}[r(t)])^2, \quad (12)$$

in order to compute the variance $\mathbb{V}[r(t)]$ of $r(t)$, it will first be necessary to compute the second statistical moment $\mathbb{E}\left[(r(t))^2\right]$. This will be done by applying the cornerstone of stochastic calculus, namely, the Itô lemma.

For the sake of completeness, now we state the Itô lemma. Let us consider the general s.d.e.
$$\begin{cases} dX(t) = f(t, X(t))dt + g(t, X(t))dB(t), \\ X(0) = X_0. \end{cases} \qquad (13)$$

Let $F(t, x)$ be a deterministic mapping so that the following partial derivatives $\frac{\partial F(t,x)}{\partial t}$, $\frac{\partial F(t,x)}{\partial x}$, and $\frac{\partial^2 F(t,x)}{\partial x^2}$ exist and are continuous. Then, $F(t, X(t))$ satisfies the following Itô s.d.e

$$dF(t, X(t)) = \left(\frac{\partial F(t,x)}{\partial t}\bigg|_{x=X(t)} + f(t,X)\frac{\partial F(t,x)}{\partial x}\bigg|_{x=X(t)} \right.$$
$$+ \frac{1}{2}(g(t,X))^2 \frac{\partial^2 F(t,x)}{\partial x^2}\bigg|_{x=X(t)} \bigg) dt \qquad (14)$$
$$+ g(t,X)\frac{\partial F(t,x)}{\partial x}\bigg|_{x=X(t)} dB(t).$$

Identifying general model (13) and (4) CIR model, one gets
$$X(t) = r(t), \quad f(t, X(t)) = \alpha(r_e - r(t)), \quad g(t, X(t)) = \sigma\sqrt{r(t)}, \quad X_0 = r_0.$$

Now we apply (14) to $F(t, x) = x^2$ with the identification $x = r \equiv r(t)$, taking into account that
$$\frac{\partial F(t,r)}{\partial t} = 0, \quad \frac{\partial F(t,r)}{\partial r} = 2r, \quad \frac{\partial^2 F(t,r)}{\partial r^2} = 2.$$

This yields
$$d(r(t))^2 = \left(2\alpha(r_e - r(t))r(t) + \sigma^2 r(t) \right) dt + 2\sigma(r(t))^{3/2} dB(t). \qquad (15)$$

Integrating on the interval $[0, t]$, one gets
$$\begin{aligned} (r(t))^2 &= (r_0)^2 + 2\alpha r_e \int_0^t r(s)ds - 2\alpha \int_0^t (r(s))^2 ds + \sigma^2 \int_0^t r(s)ds \\ &+ 2\sigma \int_0^t (r(s))^{3/2} dB(s). \end{aligned} \qquad (16)$$

Now, we take the expectation operator and then we use the properties given in (8) to obtain
$$\begin{aligned} \mathbb{E}\left[(r(t))^2 \right] &= (r_0)^2 + 2\alpha r_e \int_0^t \mathbb{E}\left[r(s) \right] ds - 2\alpha \int_0^t \mathbb{E}\left[(r(s))^2 \right] ds \\ &+ \sigma^2 \int_0^t \mathbb{E}\left[r(s) \right] ds. \end{aligned} \qquad (17)$$

Let us denote
$$m_2(t) = \mathbb{E}\left[(r(t))^2 \right], \qquad (18)$$
then (17) can be written as
$$m_2(t) = (r_0)^2 + 2\alpha r_e \int_0^t m_1(s)ds - 2\alpha \int_0^t m_2(s)ds + \sigma^2 \int_0^t m_1(s)ds, \qquad (19)$$

where $m_1(t) = \mathbb{E}\left[r(t)\right]$. Differentiating with respect to t this integral equation one sets the following initial value problem

$$\begin{cases} \dfrac{\mathrm{d}\left(m_2(t)\right)}{\mathrm{d}t} = -2\alpha m_2(t) + \left(2\alpha r_e + \sigma^2\right) m_1(t), \\ m_2(0) = (r_0)^2, \end{cases} \tag{20}$$

where $m_1(t)$ is given by (11). It can be checked that the solution to (20) is given by

$$\begin{aligned} m_2(t) &= \frac{1}{2\alpha} \mathrm{e}^{-2\alpha t} \left[2(r_0)^2\alpha - 4r_0 r_e \alpha + 2(r_e)^2 \alpha - 2r_0 \sigma^2 + r_e \sigma^2 \right. \\ &\quad + \left(4r_0 r_e \alpha - 4\alpha(r_e)^2 + 2r_0 \sigma^2 - 2r_e \sigma^2\right) \mathrm{e}^{\alpha t} \\ &\quad \left. + \left(2\alpha(r_e)^2 + r_e \sigma^2\right) \mathrm{e}^{2\alpha t}\right]. \end{aligned} \tag{21}$$

Hence, substituting expressions of $m_1(t) = \mathbb{E}\left[r(t)\right]$ and $m_2(t) = \mathbb{E}\left[(r(t))^2\right]$ given by (11) and (21), respectively, into (12), one gets the variance function of the solution stochastic process to model (4)

$$\mathbb{V}[r(t)] = \frac{\sigma^2 r_e}{2\alpha} + \frac{\sigma^2(r_0 - r_e)}{\alpha} \mathrm{e}^{-\alpha t} + \left(\frac{\sigma^2 r_e}{2\alpha} - \frac{\sigma^2 r_0}{\alpha}\right) \mathrm{e}^{-2\alpha t} \xrightarrow[t \to \infty]{} \frac{\sigma^2 r_e}{2\alpha}. \tag{22}$$

4. Parameter Estimation

After studying the model, one needs to estimate the model parameters, in this case, r_e, α, and σ from sample interest rate data. This section is addressed to face this question. As it will be seen later, we will estimate the parameters for a sample of the 1-month Euribor interest rate, although our approach is valid for other short-term interest rates. It is important to point out that, from a theoretical point of view short-term condition is required since the stochastic model (4) has been established assuming that the time increment $\Delta t > 0$ is short enough to legitimate its mathematical formulation.

Model parameter estimations will be done using two different methods. First, maximum likelihood method and, secondly, a non-parametric moment method.

4.1. Maximum Likelihood Estimation

Let $\mathcal{S} = \{r_0, r_1, \ldots, r_N\}$ be a sample of interest rates corresponding to times $t_0, t_0 + \Delta t, \ldots, t_0 + N\Delta t$, $\Delta t > 0$, respectively. Let $l(\vec{\theta}; r_0, r_1, \ldots, r_N)$ be the likelihood function associated to the sample $\mathcal{S}$, where $\vec{\theta} = (r_e, \alpha, \sigma)$ denotes the parameter vector to be estimated.

In order to obtain the transition probability density functions that will appear involved in the likelihood function, we apply the Euler-Maruyama stochastic numerical scheme to the s.d.e. (4) on the time discretization $0 = t_0 < t_1 = \Delta t < t_2 = 2\Delta t < \cdots < t_{i-1} = (i-1)\Delta t < t_i = i\Delta t < \cdots < t_N = N\Delta t$, [9, p.305]. This leads to

$$r(t_i) \approx r_i = r_{i-1} + \alpha(r_e - r_{i-1})\Delta t + \sigma\sqrt{r_{i-1}}\{W_i - W_{i-1}\}, \quad 1 \leq i \leq N, \tag{23}$$

being

$$W_i - W_{i-1} = W(t_i) - W(t_{i-1}) = W(i\Delta t) - W((i-1)\Delta t) \sim \mathrm{N}(0; \Delta t),$$

where we recall that the increments of brownian motion are gaussian with zero mean and standard deviation $\sqrt{\Delta t}$. Using the following representation for simulating the increments of brownian motion, $W_i - W_{i-1} = \sqrt{\Delta t}\eta_i$, $\eta_i \sim \mathrm{N}(0;1)$, the scheme (23) can be written as

$$r_i \approx r_{i-1} + \alpha(r_e - r_{i-1})\Delta t + \sigma\sqrt{r_{i-1}}\sqrt{\Delta t}\eta_i, \quad \eta_i \sim \mathrm{N}(0;1), \ 1 \leq i \leq N. \quad (24)$$

From this approximation one deduces the probabilistic distribution of r_i given r_{i-1}, since r_i is a linear transformation of the gaussian random variable η_i, and we have

$$r_i | r_{i-1} \sim \mathrm{N}\left(r_{i-1} + \alpha(r_e - r_{i-1})\Delta t; \sigma^2 r_{i-1}\Delta t\right), \quad 1 \leq i \leq N. \quad (25)$$

As a consequence, the transition probability density function of $r_i | r_{i-1}$ is given by

$$p(r_i | r_{i-1}; \alpha, r_e, \sigma) \approx \frac{1}{\sqrt{2\pi\sigma^2 r_{i-1}\Delta t}} e^{\frac{[r_i - (r_{i-1} + \alpha(r_e - r_{i-1})\Delta t)]^2}{\sigma^2 r_{i-1}\Delta t}}, \quad 1 \leq i \leq N. \quad (26)$$

Taking into account that the solution stochastic process to Itô-type differential equations (13) are Markov processes of order 1 [10, Th. 5.2.5, p.131] and the total probability theorem, it is easy to check that the likelihood function can be expressed as

$$l(\vec{\theta}) = p(r_0)\prod_{i=1}^{N} p(r_i | r_{i-1}; \vec{\theta}), \quad \vec{\theta} = (r_e, \alpha, \sigma). \quad (27)$$

Therefore, our goal is to maximize $l(\vec{\theta})$ to get the best parameter estimation. However, from a numerical standpoint it is more convenient, and equivalent, to handle the following objective function:

$$\min_{\vec{\theta} \in \mathbb{R}^3} L(\vec{\theta}), \quad \text{where} \quad L(\vec{\theta}) = -\ln\left(l(\vec{\theta})\right). \quad (28)$$

After replacing (26) into (27)–(28) and performing simplifications, one obtains the following expression for the log-likelihood function

$$\begin{aligned} L(r_e, \alpha, \sigma; r_0, \ldots, r_N) &= \frac{N}{2}\left(\ln(2\pi\Delta t) + 2\ln(\sigma)\right) \\ &+ \frac{1}{2}\sum_{i=1}^{N} \ln(r_{i-1}) \\ &+ \frac{1}{2\Delta t\sigma^2}\sum_{i=1}^{N} \frac{[r_i - (r_{i-1} + \alpha(r_e - r_{i-1})\Delta t)]^2}{r_{i-1}}. \end{aligned} \quad (29)$$

To estimate the model parameters (r_e, α, σ) we have considered a sample of 134 values corresponding to the 1-month Euribor interest rate from January 2nd, 2013 until July 11th, 2013 taken from [11]. These values have been plotted in Figure 1 (red points). Model parameter estimation has been computed by *Mathematica* software [12] using the Differential Evolution algorithm with time step $\Delta t = 1$ in (29). The results are collected in the second column of Table 1.

4.2. Non-Parametric Estimation

In a similar way to Subsection 4.1, let $\mathcal{S} = \{r_0, r_1, \ldots, r_N\}$ be a sample of interest rates corresponding to times $t_0, t_0 + \Delta t, \ldots, t_0 + N\Delta t$, $\Delta t > 0$, respectively. The problem lies in how to estimate the parametric vector $\vec{\theta} = (r_e, \alpha, \sigma)$. This nonparametric estimation method has been studied in [13]. In a first step, this consists of applying the Euler Maruyama scheme to the general stochastic differential equation (13) similarly as was shown in the previous subsection. Therefore, one obtain the equation (23).

After these steps, as it is mentioned in [2] we will perform the main part, which is based on the approximate expectations. One can compute this approximation by using the values of r_i, $\forall\, 0 \leq i \leq N-1$, for a discrete-time process. Obtaining the equations

$$\begin{aligned}
\mathbb{E}\left[(r_{i+1} - r_i)/\Delta t - \alpha(r_e - r_i)\right] &= O(\Delta t), \\
\mathbb{E}\left[(r_{i+1} - r_i)^2/\Delta t - \sigma^2 r_i\right] &= O(\Delta t), \\
\mathbb{E}\left[(r_{i+1} - r_i)\, r_i/\Delta t - \alpha(r_e - r_i) r_i\right] &= O(\Delta t).
\end{aligned} \quad (30)$$

Consequently, one could estimate the parametric vector $\vec{\theta}$ using the sample counterparts of the equation (30),

$$\begin{aligned}
\sum_{i=0}^{N-1} \alpha\, (r_{i+1} - r_i) &= \frac{1}{\Delta t} \sum_{i=0}^{N-1} (r_{i+1} - r_i), \\
\sum_{i=0}^{N-1} \sigma^2 r_i &= \frac{1}{\Delta t} \sum_{i=0}^{N-1} (r_{i+1} - r_i)^2, \\
\sum_{i=0}^{N-1} \alpha\, (r_{i+1} - r_i)\, r_i &= \frac{1}{\Delta t} \sum_{i=0}^{N-1} (r_{i+1} - r_i)\, r_i.
\end{aligned} \quad (31)$$

This results are collected in the third column of Table 1.

4.3. Results

In this subsection we summarize the model parameter estimations obtained using the two methods described in Subsections 4.1 and 4.2. To facilitate the comparison between both methods, the numerical results are collected in Table 1. We observe that both estimations are close. This provides reliability and robustness to our estimations.

5. Validation of the Stochastic Model

The next natural step is to verify if the parameters are good enough to be able to use them. To that end, we are going to use two different methods of measurement of goodness-of-fit, MAPE (Mean Absolute Percentage Error) and RMSE (Root Mean Square Error). Both of them will be calculated for the theoretical estimation of the values, $r(t)$.

For the parameters estimation, we will take the expected value, equation (11). Then, the values for the RMSE and MAPE measures of goodness-of-fit are show in Table 2. We can

Table 1. Model parameter estimations to the CIR model using a maximun likelihood method and a non-parameric technique considering a 1-month Euribor interest rate sample from January 2nd, 2013 until July 11th, 2013

Parameter	Maximum Likelihood Method	Nonparametric Technique
α	0.03172	0.03222
r_e	0.12043	0.12037
σ	0.00281	0.00286

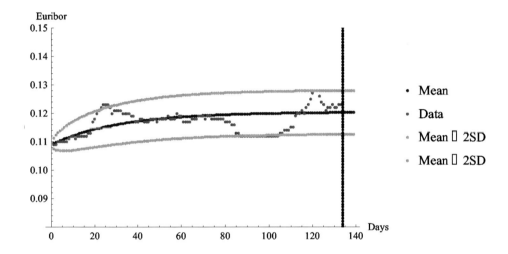

Figure 1. Model values (mean), condifence interval (mean plus/minus two standard deviations) and real data for 1-month Euribor interest rate from January 2nd, 2013 to July 11th, 2013.

see that with both methods, RMSE, and MAPE validate the model (4), and therefore, they tell us that we can take this estimation to model and predict the 1-month Euribor rates.

Table 2. Validation of the estimation parameters for the CIR model (4) by using RMSE and MAPE

Estimation Methods	RMSE	MAPE
Maximum Likelihood Technique	0.00422968	2.68099
Nonparametric Technique	0.00421973	2.67816

6. Predictions Based on the Parameters Estimation

Due to the fact that, the method have been validated, we are able to make predictions for 1-month Euribor interest rate. In Table 3 the exact value of the 1-mount Euribor, our predictions and confidence interval for the maximum likelihood estimation are shown. In the same way, but for the nonparametric estimation methods are shown in the Table 4. These confidence intervals and estimations for the 134 days have been plotted in the Figure 1, as well as, the next 5 days predictions. These latter values have been plotted after the vertical (black) line.

Table 3. Predictions for the 1-month Euribor using the likelihood estimations

Date	1-month Euribor	Punctual estimation	95% Confidence interval
12 Jul 2013	0.123	0.120268	[0.112539, 0.127997]
15 Jul 2013	0.123	0.120273	[0.112544, 0.128002]
16 Jul 2013	0.123	0.120278	[0.112548, 0.128008]
17 Jul 2013	0.122	0.120283	[0.112553, 0.128013]
18 Jul 2013	0.122	0.120287	[0.112557, 0.128018]

Table 4. Predictions for the 1-month Euribor using the nonparametric estimations

Date	1-month Euribor	Punctual estimation	95% Confidence interval
12 Jul 2013	0.123	0.120227	[0.112403, 0.128052]
15 Jul 2013	0.123	0.120232	[0.112407, 0.128058]
16 Jul 2013	0.123	0.120237	[0.112411, 0.128063]
17 Jul 2013	0.122	0.120241	[0.112416, 0.128067]
18 Jul 2013	0.122	0.120246	[0.112420, 0.128072]

Conclusion

In this paper we have shown several motivations of the so-called stochastic Cox-Ingersoll-Ross's model taking as starting point its deterministic counterpart. This model is based on an Itô type stochastic differential equation. In opposition to Vasicek's model, in CIR model, variable local volatility is considered. Nevertheless, the mean keeps having mean reversion asymptotic behaviour. Afterwards, we have provided several model parameter estimations and verified that these estimations are reliable. This has enabled us to construct good predictions since all of the real data are within of the confidence interval. Additionally, we have obtained that, the results with the CIR model are closer than the results obtained with Vasicek's model in [3].

References

[1] J.C. Cox, J.E. Ingersoll, S.A. Ross, A theory of the term structure of interest rate, *Econometrica* 53(2), (1985), 385–407.

[2] E. Allen, Modelling with Itô Stochastic Differential Equations, Springer, *Series Mathematical Modelling: Theory and applications*, 2007.

[3] J.C. Cortés, J.V. Romero, A. Sánchez-Sánchez, R.J. Villanueva, Modelling 1-month Euribor interest rate by using differential equations with uncertainty, *Applied Mathematical and Computational Science* 7(3), (2015), 37–50.

[4] O. Vasicek, An equilibrium characterization of the term structure, *Journal of Financial Economics* 5(2), (1977), 177–188.

[5] K.C. Chan, G.A. Karolyi, F.A. Longstaff, A.B. Sanders, An empirical comparison of alternative models of the short-term interest rate, *Journal of Finance* XLVII(3), (1992), 1209–1227.

[6] Y. Aït-Sahalia, Testing continuous-time models of the spot rate, *The Review of Financial Studies* 9, (1996), 385–426.

[7] M. Baxter, A. Rennie, *Financial Calculus: An Introduction to Financial Pricing*, Cambridge University Press, 1997.

[8] D. Brigo, F. Mercurio, *Interest Rate Models-Theory and Practice with Smile, Inflation and Credit*, Springer Finance, 2006.

[9] P.E. Kloeden, E. Platen, Numerical Solution of Stochastic Differential Equations, Springer, *Series Applications of Mathematics: Stochastic Modelling and Applied Probability* 23, 1992.

[10] T.T. Soong, Random Differential Equations in Science and Engineering, Academic Press, *Series Mathematics in Science and Engineering* 103, 1973.

[11] www.emmi-benchmarks.eu/euribor-org/euribor-rates.html

[12] https://www.wolfram.com/mathematica/

[13] Stanton R., A nonparametric model of term structure dynamics and the market price of interest rate risk, *Journal of Finance* LII, (1997), 1973–2002.

[14] Y. Maghsoodi, Solution of the extended CIR term structure and bond option valuation, *Mathematical Finance* 6(1), (1996), 89–109.

[15] T.A. Marsh, E.R. Rosenfeld, Stochastic processes for interest rates and equilibrium bond prices, *Journal of Finance*, 38(2), (1982), 635–646.

In: Modeling Human Behavior: Individuals and Organizations   ISBN: 978-1-53610-197-3
Editors: L. Jódar Sánchez, E. de la Poza Plaza et al.   © 2017 Nova Science Publishers, Inc.

Chapter 14

CONSUMERS' MULTI-HOMING OR MULTIPLE DEMAND

Cristina Pardo-García[1,*] *and*
María Caballer-Tarazona[1]
[1]Department of Applied Economics, Universitat de València,
València, Spain

ABSTRACT

A multi-homing or multiple purchase behavior by consumers is arising in some markets of apparently substitute goods. A consumer can do multiple consumption in order to be able to choose provider in the future. This behavior usually appears in industries where goods are acquired in two times, that is, paying a membership fee and a price per service afterwards. Are firms pleased or harmed by the existence of multi-homing? Firms find profitable if consumers multi-home, since they are doubling demand.

Moreover, competition is softened, a price reduction make consumers change from single-homing to multi-homing, so firms are not stealing market share to their rivals but increasing demand through multi-homing behaviour.

Keywords: multi-homing, multiple demand, two-part tariff

INTRODUCTION

The received literature in the address approach to product differentiation has considered so far that consumers buy at most one unit when they are choosing between substitute goods. Our purpose in this chapter is to move away from this restrictive assumption and allow consumers to buy from more than one firm in case of substitute goods. Sometimes an election dilemma appears when people have to choose one good or the other. They might find useful not to be tied to one product in order to be able to decide afterwards between all possibilities

---

[*] Corresponding Author Email: cristina.pardo-garcia@uv.es.

the market offers. In case this decision is taken before the real necessity arises, it is possible that consumers do not really know whether to take one or the other good, or they just do not want to be committed to a given product at that moment[1]. Then, a safety choice is to buy both substitute products leading to a kind of consumer behavior which can be denoted by multi-homing behavior. Then, substitute goods are purchased together like complements. Besides, consumers might perceive goods as different from one another, because of different characteristics even though both goods seem to satisfy, a priori, the same necessity[2]. What multi-homing implies for buyers is the possibility of further multiple purchase. This behavior is showing strong preferences for diversity through not limiting their election set, keeping their right to choose from different firm's products in the future.

There are examples of multi-homing behavior in totally different markets. Some of them imply a complementary use of a private service in an industry with a free public service. Health care is free in many countries with a National Health System, where everyone is covered. However, this public provision is usually crowded and therefore, some patients find useful to subscribe and pay for private services. They are thus able to choose in case of illness in which one of the health services they prefer to be treated on. Another example is the media industry where usually public TV is free, but again we find some people that subscribe to a pay-per-view TV platform increasing their election set and choosing later from which provider they watch, depending on the programmes they offer. Obviously, when we are referring to free public services, that means that the price is zero at the point of consumption, but health services and public TV are financed through taxes paid by citizens. However, they cannot avoid paying through their taxes for these services.

Multi-homing behavior is also arising in industries with two or more competing private firms. For instance, in the video game market, video game lovers usually hold both the PS3 console as well as the X-box one. Then, if a specific game is exclusive to one platform, multi-homing consumers are able to play since they have both consoles. If the game is sold in all platforms, they always have the option to choose the one with lower price. People also do multi-homing in banking services since some customers have bank accounts in different companies. Banks usually offer the same services, so multi-homing must imply some advantages. The first one is protection against bankruptcy, if a bank goes bankrupt and the State guarantees only up to certain amount of money, it is better to have your savings split in different firms. The second is using a wider range of ATM's, when banks have different systems of debit and credit cards (as Visa or Master Card). The third one, users can take profit of different services as loans or fixed term deposits by choosing the entity which has better conditions at that moment. Note that it is a usual practice for banks to discriminate prices depending on the identity of the customer, that is, whether it is a customer coming from another bank or an own one.

A *two-part tariff* pricing structure is commonly implemented in those examples above. That is, customers have to pay both a subscription fee or membership rate and a price per service. The subscription fee can be understood as just a *membership fee* or the price of the

---

[1] Armstrong (2006) states that in dynamic price discrimination "consumers might not know their preferences for future consumption at the time of their initial dealings with a firm".

[2] As in Anderson and Neven (1989): "Hence the type of product differentiation considered is such that there is no consensus among consumers over some product quality. Rather, we try to describe a market segment in which products are essentially comparable but differently valued by consumers. Hence, we deal here with pure horizontal differentiation".

product needed for the satisfaction of the future consumer's needs[3]. For instance the annual fee in a health insurance, or the price of a specific device of a pay-per-view TV platform, or the price of the video game console. Then, a *price per service* is charged to the users in an aftermarket or aftermath period. For example video game users pay for every game they want, pay-per-view TV consumers have to buy programmes from certain TV channels. Different pricing strategies can lead to other cases: linear pricing, in which subscription fee is equal to 0; or flat fees, in which the price per service is equal to 0.

Users' heterogeneity is an important feature to understand multi-homing behavior. It gives rise to different types of consumers behavior. On the one hand, there are groups of consumers who are loyal to one of the firms. On the other hand, there will arise consumers who are better off if they buy from both firms. This latter type accept to pay the subscription fee for both firms in order to be able to demand from any one in the future or even from both. Think about a patient that pays the membership fee and then, in the aftermath period, demand a flu treatment to the private provider while a heart operation to the public provider. In the TV industry, a consumer can have public and pay-per-view TV platforms, but one day he can choose to watch a football match in the public service while the next day he can pay to watch a film in the pay television.

Thus, this *double* purchase of substitute goods must pay off: the gains they obtain outperform prices. As already said, multi-homing consumers are indeed considering substitute goods as if they complement each other. Goods that were non-accessible ex-ante, as physicians (or treatments) in different health companies or exclusive games for different type of consoles, are now reachable with this *doubled* subscription or purchase behavior. Several questions will easily arise: Is multi-homing desirable for firms? Is competition intensified or softened?

Think of a duopoly model where firms are selling a differentiated good and consumers are heterogeneous in their preferences for that good. If a consumer wants to consume the good of a particular firm, it has to pay an initial fee which enables her to further buy the service or product offered by that duopolist. Since each consumer envisions aftermarket consumption, its initial decision on which product to buy takes into account aftermarket prices, so her willingness to pay for each product is positively correlated with aftermarket consumer surplus. Users can either buy one of the products, becoming exclusive or single-homing consumers, or buy both, so they *double purchase* or they do multi-homing. In the case a user buys both products, she is in a better position with respect to aftermarket consumption, since she can choose among the products offered by each duopolist. In that way, a single-homing consumer is expecting to get a lower consumer surplus than the one it would obtain in case of multi-homing. Therefore, a multi-homing behavior would be observed if the extra price paid for the second differentiated good is offset by the increase in aftermarket consumer surplus. Similarly, there are two opposing effects on the production side. With single-homing, each firm can behave as a monopolist of its patronized users, thus getting a higher aftermarket profit per capita. When multi-homing is at place, firms exchange a higher profit per capita as individual demand is more elastic, for greater aggregated demand since multi-homing implies an increase in total demand. When the second effect dominates the former, multi-homing might be observed at equilibrium.

---

[3] There are informational costs people also have. They should be informed about the different characteristics of the goods or services each firm provides to decide which one is the most desirable to use in a certain moment.

The Related Literature

Most of the existing literature has studied multi-homing in a two-sided market context, because it is frequently observed in industries such as software, payment systems or video game markets. Examples of this literature are Rochet and Tirole (2003), Gabszewicz and Wauthy (2005) or Choi (2010). In a two-sided markets context, Hagiu (2006) analyzes a model in which only sellers can do multi-homing and platforms charge fixed fees and variable fees (royalties) on each buyer-seller transaction. Platforms distinguish between free and captive buyers and can price discriminate in access fees but the same royalties are charged to all buyers and sellers. Hagiu (2006) concludes that sellers' multi-homing behavior relaxes platform competition.

However, multi-homing behavior does not need a two-sided market context as a *sine qua non* condition to appear. We want to focus on this characteristic of consumers buying from two competing firms assuming away a two-sided market context. Only few papers have undertaken this approach. Anderson and Neven (1989) designed a model of combinable goods where only linear pricing is considered. Equilibrium allocations for firms correspond to maximum differentiation, which is also the social optimum. All consumers mix goods, obtaining their most preferred variety. Hoernig and Valletti (2007) present a model of combinable goods. Consumers buy at most one unit of the good, but they can mix, buying partially to each firm to form this unit. Their model is a two-firm Hotelling setting inspired in the media industry. First firms choose locations and then offer two-part tariffs composed by subscription fees and pay-per-view fees. At equilibrium there are consumers who mix combinable goods[4] choosing a share from different goods in the market to compose one unit of the good. The two-part tariff pricing strategy is compared to linear pricing and flat fees. They found that it is not profit maximizing to set a free subscription fee nor subsidizing subscription, therefore, equilibrium subscription fees are always positive.

Another approach to multi-homing is Kim and Serfes (2006) multiple purchase model. Those authors analyze a location duopoly model where consumers are allowed to make multiple purchases, they can buy one unit of every firm in the market. It is assumed that the marginal utility of a second purchase is lower than the utility of the first purchase. They found that when the added utility of buying a second good does not reach certain level, the equilibrium is the same as in the Hotelling model, half of the consumers buy only from one firm and the other half buy only from the other firm. However, if this added utility reaches certain level a new group of consumers buying both brands appears.

If that is the case, price competition changes since a decrease in one firm's price implies an increase in its demand, but it does not happen through the stealing of other firm's demand. Next, firms are allowed to choose their locations. Assuming quadratic costs, if the added utility reaches certain level, the Hotelling's principle of minimum differentiation is restated. The reason for the firms to locate closer to each other is an aggregate demand creation effect. The increase in willingness to pay due to double purchase is not an arbitrary constant but it is a function of the larger availability and consumption users are expecting in the aftermarket, which, in turn, depends on aftermarket prices. This links both elements in the two-part tariff

---

[4] Hoernig and Valletti (2007): "Combinable goods are substitutes for customers that decide to buy exclusively, but are complements for customers that decide to mix, as their combination matches more closely consumers' ideal goods".

pricing in a natural way. Since multi-homing users are able to choose from the product sold by both duopolists they will achieve a higher surplus, as they have a more elastic aftermath demand.

Multiple purchase is also related to the *aftermarkets* literature. Aftermarkets are complementary goods or services for a durable good such as maintenance, repair services, upgrades or replacement parts. The producer sells its good to consumers who are someway *locked-in* in the next period when a complementary service is needed, since consumers prefer to purchase the maintenance service to the same provider. Kaserman (2007) design a model supporting efficiency-based theories to engage in tying in the aftermarket.

In these models, durable-goods producers have incentives to monopolize the maintenance market. One expects this practice to be harmful, although this can be false for different reasons.

Morita and Waldman (2004) show that, as happens with leasing, monopolizing the maintenance market is another way to eliminate time inconsistency problems. It lets the producer to internalize the effects of its behavior in the units previously produced. Carlton and Waldman (2010) study the monopolization of three different competitive aftermarkets: just maintenance, upgrades or remanufactured parts. They find that, although being counter intuitive, in many cases welfare is not reduced because it leads to reduce inefficiencies in the aftermarket. These inefficiencies can come from switching costs, cross-subsidization of consumers who upgrade by consumers who buy new units in the second period, or higher cost for worn out parts to be remanufactured.

METHODS

An Approach to Multi-Homing Demand

Consider a market for a horizontally differentiated product which is served by a duopoly. In order to buy the good from firm i, where $i=A, B$, users have to pay a membership rate first, denoted by r_i; and then pay a price p_i for the product offered by that firm, e.g., the aftermarket or aftermath price. Users are allowed to buy both products then paying two membership rates and two aftermarket prices. In the event users only buy from one firm, this will be denoted as a single-homing behavior; while buying from both firms will be named a multi-homing behavior. Users are heterogeneous in their preferences with respect to each differentiated good and view them as imperfect substitutes. In order to account for such heterogeneity, we consider there is a continuum of users uniformly distributed with density one in the product space, the interval [0, 1]. For simplicity, we assume the duopolists are located at a different extreme point of the product space, that is, the product of firm A is located at zero, while that of firm B is placed at one.

As already mentioned, consumers can either decide to pay the membership rate of one of the products, both or none, once they observe the pair of membership rates issued by the duopolists. In case a consumer decides to pay only one membership rate, the chosen provider is a monopolist for this consumer in the aftermath period, as she only has the option to demand from this firm. However, when a consumer selects multi-homing, she has an individual demand for each of the products. Providers are duopolists for this multi-homing consumer, since she is enabled to choose among the two products sold in the market. A consumer can pay both membership fees and also both aftermath prices.

Regarding the maximum willingness to pay for product i, or in other words the product i's reservation price, is denoted by v_i; while the maximum willingness to pay for both products is denoted by v_{AB}. We assume that $v_{AB} > max\{v_A, v_B\}$, since the maximum willingness to pay for both products should be higher than the maximum willingness to pay for any of the individual products, A or B. Indirect utility of user located at point l, can take different expressions depending on which firm he/she patronizes:

- $v_A - r_A - t\alpha_{lA}$ if only patronizes firm A
- $v_B - r_B - t\alpha_{lB}$ if only patronizes firm B
- $v_{AB} - r_A - r_B - t(\alpha_{lA} + \alpha_{lB})$ if patronizes both firms

Obviously a fourth possibility is that user located at point l does not sign in for any product, a less interesting case. We are going to rule out this option to focus on the single versus multi-homing decision.

Parameter t measures the per unit of distance disutility of a user not consuming her most preferred good and higher t implies more differentiation in the market. Also α_{li} denotes the distance from the location point of consumer l to the location of product i, satisfying that $\alpha_{lA} + \alpha_{lB} = 1$, since from each point in the interval [0, 1] the distance to get both products, A (located at 0) and B (located at 1), is 1. Note that when a consumer chooses multi-homing has to incur on the disutility of visiting both locations and therefore this disutility is always equal to one regardless of the consumer address or location point. As a consequence, the increase in disutility derived from multi-homing is higher for consumers located at the interval ends. Therefore, the users with multi-homing behavior are more likely to appear in the middle of the interval [0, 1].

This partitions the characteristics space into several segments (possibly three) or market shares: the segment of consumers patronizing firm A, denoted by s_A, those patronizing firm B, denoted by s_B and, if they arise, those users that choose multi-homing and therefore are potential clients of both firms, denoted by s_{AB}, (see Figure 1). Note that in general each firm has two sources of revenue, the membership rates and the actual selling of the good and also each firm has two different types of consumers, each type demanding different amounts of the good for a given price, which is the same for both types of consumers.

Each consumer in the interval [0, 1] chooses to buy from the firm that implies a higher indirect utility for himself, which is represented by the thicker solid line. Each firm i has therefore a total number of users that equals $s_i + s_{AB}$, but there are two types of consumers in terms of their profitability for the firm. Multi-homing would arise if membership fees are not too large. Providers can induce a scenario with only single-homing consumers by fixing high enough membership fees since consumers will only pay one membership fee. That would be the case if the horizontal line representing the multi-homing behavior was lower than the point in which both diagonal solid lines (consumers' utilities when buying from firms A or B) cross, in the center of the graph. Similarly, for low enough membership fees all users can become multi-homing ones. That would be the case if the horizontal line representing the multi-homing behavior was higher than the points in which diagonal solid lines, consumers' utilities when buying from firms A or B, cross with the left and right vertical axes, respectively.

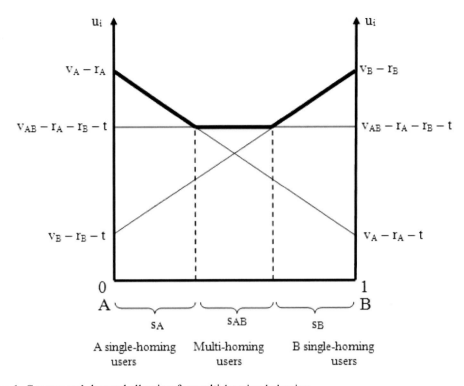

Figure 1. Consumers' demand allowing for multi-homing behavior.

Interestingly enough those users with multi-homing behavior are located in the center of the characteristics space and therefore there is no direct contact between both types of single-homing users. Once multi-homing is at place, competition is not as fierce as in a standard case with only single-homing users. A decrease in the rival's membership fee implies a lower profitability but the firm still retains the same number of consumers, with single-homing behavior the same decrease would imply the loss of users.

Both prices charged by the same firm should move in opposite directions. If a firm increases the membership fee, the aftermath price will be decreased, and vice versa. Firms have two instruments to extract consumer surplus, so if they charge a higher price on one side, they must lower the other price, in order to compensate somehow consumers and retain them. A consumer will be willing to pay a higher membership fee only if prices in the future will be lower, to compensate the loss in consumer surplus with a future gain in the subsequent aftermath period. We observe this pattern in many examples in real life. If there is a structure of subscription fee and then a price per service, firms can overcharge one of them and leave free the other one. An amusement park can charge a higher subscription fee for one year and then offer free unlimited access to the customer in that period. In the TV industry, a device can be given for free but then users have to pay for any TV programme. In printers and cartridges a similar substitution is found, printers are very cheap while cartridges have an expensive price.

CONCLUSION

A lower aftermarket price is making more attractive to consumers to subscribe a particular firm. Since consumers are more willing to pay the membership fees when they anticipate low aftermath prices, firms reduce aftermath prices to enlarge the share of multi-homing users. The reason is that this expansion effect in demand more than compensates the reduction in per capita profits.

Duopolists face two types of consumers, exclusive and multi-homing consumers. Obviously, this latter group has a more elastic demand, they are able to pay a lower price since they can choose provider. As the multi-homing set of consumers appears in the center of the characteristics space, between both sets of exclusive consumers, competition between firms is softened. Users can switch from the firm i's single-homing users group to the multi-homing group. A decrease in the rival firm's price means that some own exclusive consumers can now become multi-homing users. Therefore, that firm is losing the exclusivity over some consumers, but it is still selling to them. Multi-homing is thus softening firms' competition.

REFERENCES

Anderson, S. P. and Neven, D. J. (1989). Market efficiency with combinable products. *European Economic Review*, 33 (4), 707-719.

Armstrong, M., (2006). Recent developments in the economics of price discrimination. In: Blundell, R., Newey, W., Persson, T. (Eds.), Advances in Economics and Econometrics: Theory and Applications: Ninth World Congress. Cambridge University Press, Cambridge.

Carlton, D. W., and Waldman, M. (2010). Competition, monopoly and aftermarkets. *Journal of Law, Economics and Organization*, 26 (1), 54-91.

Choi, J. P. (2010). Tying in two-sided markets with multi-homing. *Journal of Industrial Economics*, 58 (3), 607-626.

Gabszewicz, J. J. and Wauthy, X. Y. (2005). Markets with cross network externalities as vertically differentiated markets. Working paper, November 2005.

Hagiu, A. (2006). Pricing and commitment by two-sided platforms. *RAND Journal of Economics*, 37 (3), 720-737.

Hoernig, S. H. and Valleti, T. M. (2007). Mixing goods with two-part tariffs. *European Economic Review*, 51 (7), 1733-1750.

Kaserman, D. L., (2007). Efficient durable good pricing and aftermarket tie-in sales. *Economic Inquiry*, 45 (3), 533-537.

Kim, H., and Serfes, K. (2006). A location model with preference for variety. *Journal of Industrial Economics*, 54 (4), 569-595.

Morita, H. and Waldman, M. (2004). Durable goods, monopoly maintenance, and time inconsistency. *Journal of Economics and Management Strategy*, 13 (2), 273-302.

Rochet, J.-C. and Tirole, J. (2003). Platform competition in two-sided markets. *Journal of the European Economic Association*, 1 (4), 990-1029.

In: Modeling Human Behavior: Individuals and Organizations   ISBN: 978-1-53610-197-3
Editors: L. Jódar Sánchez, E. de la Poza Plaza et al.   © 2017 Nova Science Publishers, Inc.

Chapter 15

MODELLING LEARNING UNDER RANDOM CONDITIONS WITH CELLULAR AUTOMATA

L. Acedo[*]
Instituto Universitario de Matemática Multidisciplinar,
Universitat Politècnica de València, Spain

Abstract

In this work we use the Hebbian cellular automata model to study the learning process under random conditions, i.e., when only a fraction of the pattern is available for learning at every time-step. We will show that a phase transition in the learning process takes place allowing for the complete learning of the patterns if the fraction, p, of the given pattern available at every time-step is greater than $p > 0.5$ and the number of neurons is sufficiently large. The consequences for learning in the real world, where random conditions are ubiquitous, and the possibility for a determination of this threshold by experimental psychology techniques are also discussed.

PACS: 05.45-a, 52.35.Mw, 96.50.Fm

Keywords: learning modeling, Hebbian imprinting, cellular automata

AMS Subject Classification: 60H30, 34F05, 62P20, 91,B25, 91B70

1. Introduction

Learning processes are, perhaps, one of the most studied capacities of the brain. During the past century, it was established by neurophysiological research that memories are stored in the neurons synapses by a reinforcement mechanism called Long Term Potentiation (LTP) [1]. This synaptic plasticity was anticipated by Donald O. Hebb from the perspective of psychology and it has become a paradigm in the modern theory of learning [2]. This Hebbian principle can be implemented in cellular automata models if neurons are conceived

---

[*]E-mail address: luiacrod@imm.upv.es

as individual automatons with binary states and a weight, J_{ij}, is associated to every bond connecting two neurons. For a given set of weights the cellular automata network evolves following the well-known Glauber dynamics of the Ising model at zero temperature [3]. However, the weights are not constants, but under the presentation of given pattern they update following a slower dynamics by increasing or decreasing according to the product of the states of the neurons i and j at the previous time-step. This is the mathematical analogy of Hebbian imprinting to be formulated as follows:

$$J_{ij}(t) = J_{ij}(t-1) + cs_i(t-1)s_j(t-1) , \qquad (1)$$

where $s_i(t-1)$, $s_j(t-1)$ are the states of the neurons i, j at time $t-1$ both of them connected by the bond ij and c is a small constant.

These models are known as Attractor Neural Networks or Hopfield networks as they were proposed by this author in 1982 [4]. By using only neuron automatons with two states: 0 for quiescent and 1 for firing these networks are capable of exhibiting associative memory, i.e., to recall a stored pattern from another pattern with includes only a fraction of the original. From another point of view, the acquisition of associated memories (between colors, sounds or any other sensory stimuli) is thought to be the result of the strengthening of connections between neurons or assemblies of neurons that represent the associated memories [5].

Ethology research and neuroscience have discovered that learning is also present in insects, specially in the form of association of olfactory stimuli and odour discrimination [6]. But in a natural environment the ability to learn and recall patterns must face the difficulty posed by the fact that these patterns are rarely perfectly and repetitively presented to the subject as required by the standard attractor neural network model. But, even under this restrictive conditions, learning of the whole pattern is achieved as it is deduced from our own experience and experimental psychology.

In this paper we study this learning process under imperfect presentation of the patterns in the context of attraction neural networks with boolean cellular automata. We find that generalization to the complete pattern is achieved if, at least, a random fraction with a fifty per cent of the pattern is used as stimulus to activate the network every time-step during the training stage. Consequently, this random learning process is characterized by a phase transition behaviour with a threshold $p = 0.5$ for the random fraction of the pattern presented as stimulus. Below this threshold, the pattern is not adequately stored in the neural network after the training stage but it is almost exactly stored for $p \geq 0.5$. This result suggests that some insight on the performance of test subjects as learners of patterns from which only a fraction is shown in a given time-scale could be gained from the theoretical model.

2. Methods

Our model is the standard attractor neural network with N nodes and $N(N-1)/2$ bonds connecting them. Each node, i, contains a Boolean automaton with two possible states: $s_i = +1$ (representing firing neurons) and $s_i = -1$ (quiescent neurons). Each bond ij is characterized by a strength parameter J_{ij}, $i = 1, \ldots, N$, $j = 1, \ldots, N$ for $i \neq j$. Self-interaction is not allowed so $J_{ii} = 0$ by definition. Moreover, the strength of the interaction going from neuron i towards neuron j is considered the same as its reciprocal from j to i and, consequently, the matrix $J_{ij} = J_{ji}$ is symmetric.

For the purpose of the visual interpretation of the patterns we consider that the neurons are arranged in a two-dimensional array although all pairs ij are connected forming a complete graph topology. The system evolves in two stages:

- The network receives an external two-dimensional stimulus (complete or incomplete) which can be codified in binary code. For example, a black and white image. This stimulus activates the neurons in such a way that neurons receiving information about the white pixels of the figure start firing whereas the other remain quiescent. During this presentation of the stimulus the synaptic strength matrix, J_{ij}, change according to the imprinting rule in Eq. (1) with a small parameter c.

  We call this a training stage. This kind of representation is a simple implementation of information processing in the visual cortex as it is studied in experimental neuroscience [7].

- After the training stage we study the retrieval of the full pattern from a random initial condition of the network by using Glauber dynamics [3] as follows:

$$s_i(t+1) = \text{sign}\left(\sum_{j=1}^{N} J_{ij} s_j(t)\right) \quad i = 1, \ldots, N, \qquad (2)$$

where the sum within the brackets corresponds to the field acting upon neuron i as a consequence of the rest of the network. For positive strength parameters J_{ij} the firing neurons induce their neighbours to start firing at the next time-step.

In the next section we will discuss the simulation results for the imprinting of square and circular patterns presented to the neural network only with a fraction p at each time-step.

2.1. Imprinting of Patterns with Randomness

We consider a training stage for a given pattern. In the two-dimensional lattice these patterns should correspond to a square or a circle. During the training stage they are presented 100 times to the network but in such a way that only a random fraction $p < 1$ is received correctly by the neural network. Our objective is to study the effect of this noise in the learning process by determining a measure of the quality of the retrieval in comparison with the complete pattern we intended to imprint. This is usually done in terms of the Hamming's distance between the original pattern $\mathbf{s}$ and the retrieved one $\mathbf{s}'$:

$$d(\mathbf{s}, \mathbf{s}') = \frac{N}{2} - \sum_{i=1}^{N} s_i s_i', \qquad (3)$$

where the sum extends over the whole network. In Figures (1) and (2) we show the evolution of Hamming's distance during the retrieval phase for circular and square patterns imprinted with different degrees of randomness. We notice that for $p = 0.8$ the pattern is imprinted almost perfectly in the network because the Hamming distance tends to zero with

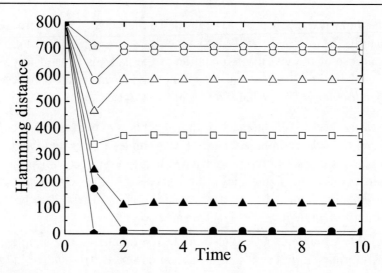

Figure 1. Hamming distance from a circular pattern learned with different degrees of randomness during the retrieval stage as a function of the time-step for $p = 0.2$ (pentagons), $p = 0.4$ (circles), $p = 0.45$ (triangles), $p = 0.5$ (squares), $p = 0.55$ (closed triangles), $p = 0.6$ (closed circles) and $p = 0.8$ (closed pentagons). The training parameter was $c = 0.001$ and 100 training time-steps were used. The size of the network was $N = 400$.

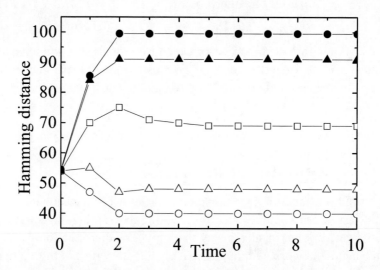

Figure 2. The same as Figure 1 but for a square pattern for $p = 0.4$ (closed circles), $p = 0.45$ (closed triangles), $p = 0.5$ (squares), $p = 0.55$ (triangles), $p = 0.6$ (circles).

the evolution time from a random initial condition. However, pattern retrieval deteriorates faster if the imprinting was carried out with imperfect patterns.

A particular case is displayed in Figures 3 and 4 for a rectangular pattern imprinted under random conditions. After the imprinting the network is prepared in a totally random state and evolution takes place according to Glauber dynamics. We notice that the rectangular pattern is retrieved on one side of the network but the other half still exhibits some noise.

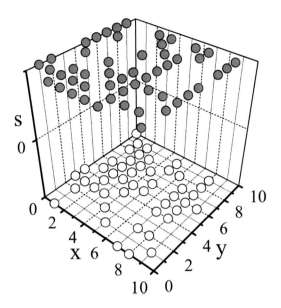

Figure 3. Random initial state in a neural network with $N = 100$ nodes. Firing networks are displayed as grey circles, resting neurons as open circles. Notice that, although the neurons are arranged in a two-dimensional lattice, the graph of connections among them is complete.

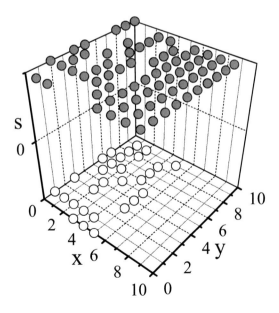

Figure 4. The same as Figure 3 after the evolution using Glauber dynamics in Eq. (2). A rectangular pattern is retrieved but the retrieval is not perfect because some noise persist on the other side of the network.

2.2. Phase Transition

In the last section we have shown that patterns which have been imprinted in such a way that a random proportion $p > 0.5$ of the pattern is presented ("observed") by the network simultaneously are susceptible of being retrieved perfectly or almost perfectly. But for $p < 0.5$ retrieval is very poor. This suggest that a kind of phase transition with a threshold $p = 0.5$ appears in this learning experiment. To analyze it more deeply we have studied the efficiency of pattern's learning in terms of $F(p)$ defined as the average of the Hamming's distance from the retrieved pattern to the real pattern. This average is performed over many retrieving stages starting from a random pattern after a learning stage in which only a random proportion p of the pattern is presented allowing for the modification of synaptic strengths following Eq. (1) in T steps.

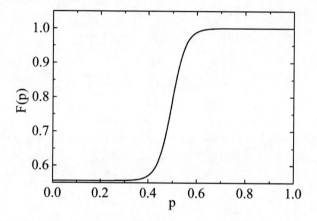

Figure 5. Average retrieved fraction of an ideal pattern, $F(p)$, vs the random proportion of the image shown to the network during the imprinting stage, p. Imprinting was performed with $c = 0.001$ during $T = 100$ steps. A network with $N = 3600$ neurons was used.

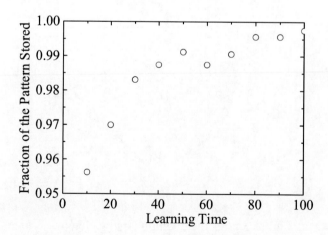

Figure 6. Fraction of the pattern retrieved vs learning time for $p > 0.5$. Notice that retrieval is very good even for 20 steps of exposure to the pattern.

The result for the average retrieving function $F(p)$ is shown in Figure 5 for a network with $N = 3600$ neurons. We find that a characteristic S-shaped function describes the transition and, consequently, the proportion of the retrieved pattern increase sharply for $p > 0.5$. This could have important consequences for the learning process in stochastic environments as it clearly indicates that neural networks are fairly robust devices for storing information about the world. Another question we can pose with this model is the influence of the number of times the pattern stimulates the network on the fraction of the pattern stored for $p > 0.5$. The influence of the learning time is analyzed in Figure 6 where we can see that, for more than 50 expositions to the pattern during the learning stage, the retrieval fraction is almost constant and larger than a 99%.

Conclusion

A model for learning of visual patterns in random conditions has been discussed. Our model is based on the standard Hopfield attractor neural network with synaptic reinforcement. A learning stage is simulated by the activation of a certain number of nodes in the network corresponding to the spatial distribution of a given pattern (a circle or a square) but in such a way that only a fraction of the pattern is activating the network simultaneously.

This fraction, p, is a measure of the randomness of the environmental conditions. In a second phase we apply Glauber dynamics from an unprepared initial condition to determine to what extend the whole pattern is retrieved. We show that, despite the handicap posed by partial presentation of the pattern to be imprinted, almost perfect retrieval is achieved for $p > 0.5$ but very poor or no retrieval at all is obtained for $p < 0.5$.

In we interpret the retrieval percentage for the imprinted pattern as an order parameter in terms of the probability p of activation of a given neuron in the network during the learning stage, the phase transition in the learning process can be characterized as first-order and the discontinuity in the order parameter is characteristic in other phenomena in networks such as discrete percolation [8]. It is expected that the transition becomes sharper and sharper as the number of neurons (nodes) in the network increases as well as the size of the pattern.

The main result of this work concerns the capacity of neural networks to learn a whole pattern by assembling the partial information received at any given instant. We have shown that pattern learning (and even perfect learning) is possible if, at least, a fifty per cent of the information is presented simultaneously during the learning stage. However, if less than a fifty per cent of the information excites the corresponding spatial distribution of neurons in the network we get no retrieval at all. This could have important implications in the design of learning strategies. As an example, Holford et al. studied the use of stereoscopic viewing for learning spatial distributions [9] because it allows for the 3D viewing of the structure and, consequently, sends more information simultaneously to the visual cortex.

Associative learning is also studied in animals [10, 11] and they could provide also a test case for studying the effect of randomness in the learning process. It is also known that even insects are capable of conditional associations and, particularly, in relation to colour [12, 13, 14]. It would be interesting if conditioning experiments with random conditions could be tested with animals to assess the degree of learning and to check the reliability of our neural network model.

Finally, we could say that even simple neural models as the Hopfield network are still useful to study a repertoire of conditions for learning and pattern retrieval and to make concrete predictions susceptible of being tested with real neural networks. Consequently, these mathematical models suggest us new ways of designing experiments in psychology and animal behaviour and it would be interesting if, actually performed, they confirm the existence of a learning phase transition in random conditions.

References

[1] Paulsen, O. and Sejnowski, F. (2000). Natural patterns of activity and long-term synaptic plasticity, *Current Opinion in Neurobiology*, 10(2), 172-179.

[2] Hebb, D. O. (1949). *The Organization of Behaviour*, Wiley, New York, U.S.A..

[3] Bar-Yam, Y. (1997). *Dynamics of Complex Systems*, Addison-Wesley, Reading, MA, U.S.A..

[4] Hopfield, J. J. (1982). *Neural networks and physical systems with emergent collective computational abilities*, Proceedings of the National Academy of Sciences of the USA, 79(8), 2554-2558.

[5] Kohonen, T. (1989). *Self-Organization and Associative Memory*, Springer-Verlag, Berlin, Germany.

[6] Parnas, M., Lin, A. C., Huetteroth, W., Miesenböck, G. (2013). *Odor discrimination in Drosophila: from neural population codes to behavior*, Neuron, 79(5), 932-944.

[7] Cox, D. D., Savoy, R. L. (2003). Functional magnetic resonance imaging (fMRI) "brain reading": detecting and classifying distributed patterns of fMRI activity in human visual cortex, *NeuroImage*, 19, 261-270.

[8] Grimmett, G. (1999). *Percolation*, A Series of Comprehensive Studies in Mathematics, Second Edition, Springer-Verlag, Berlin, Germany.

[9] Holford, D. G., Kempa, R. F. (1970), The effectiveness of stereoscopic viewing in the learning of spatial relationships in structural chemistry, *Journal of Research in Science Teaching*, 7, 265-270.

[10] Zentall, T. R., Wassermann, E. A., Urcuioli, P. J. (2014). *Associative Concept Learning in Animals*, Department of Psychological Sciences Faculty Publications, Purdue University, Paper 67.
http://docs.lib.purdue.edu/cgi/viewcontent.cgi?article=1070&context=psychpubs
(Accessed March, 8th, 2016).

[11] Fagot, J., Kruschke, J. K., Dépy, D., Vauclair, J. (1998). Associative learning in baboons (Papio papio) and humans (Homo sapiens): species differences in learned attention to visual features, *Animal Cognition*, 1, 123-133.

[12] Weiss, M. R. (1995). Associative colour learning in a nymphalid butterfly, *Ecological Entomology*, 20, 298-301.

[13] Kelber, A. (1996). Colour learning in the hawkmoth Macroglossum stellatarum, *The Journal of Experimental Biology*, 199, 1127-1131.

[14] Weiss, M. R., Papaj, D. R. (2003). Colour learning in two behavioural contexts: how much can a butterfly keep in mind ?, *Animal Behaviour*, 65, 425-434.

Chapter 16

CAPTURING THE SUBJACENT RISK OF DEATH FROM A POPULATION: THE WAVELET APPROXIMATION

I. Baeza-Sampere * *and F. G. Morillas-Jurado*
Department of Applied Economics.
University of Valencia, Valencia, Spain

ABSTRACT

In the demographic and actuarial field, widely used tools are the life tables or mortality tables. This instrument is used to summarize the experience of mortality observed in a region or period. The information shown in these tables correspond to the called biometric functions. Among these biometric functions can be found: the population at risk of dying, E_x; the number of persons who survey or die at age x, l_x o d_x respectively; or the probability of dying between ages x to x + 1, q_x. This work focuses its efforts in making estimates of the underlying values of $\{q_x\}$; $x = 0;\ldots; \omega$ (ω being the highest age considered in the study), which can be seen as a realization of a time series. The random nature of the phenomenon of study and the impossibility of its replication makes that the values of biometric functions to be unknown that only estimations can be obtained. Then, to estimate underlying mortality rates from the observed rates it is common to use graduation techniques. So, you can find parametric techniques using adjustments to certain families of exponential functions (Heligman and Pollard, 1980); semi-parametric techniques using splines bases (Forfar, McCutcheon and Wilkie, 1988); or nonparametric techniques as Kernel graduation (Gavin, Haberman and Verrall, 1993) or the wavelet (Baeza and Morillas, 2011) graduation.

This study focuses on wavelets techniques and the objective is to generalize the results obtained in (Baeza and Morillas, 2011). Thus, a methodology is presented in order to graduate the mortality observed over the entire range of ages (from birth to the age limit considered).

It is known that mortality has a nonlinear behavior and that it varies by age segments, therefore, an important disadvantage to perform the wavelet graduation is the amount of data available: normally you have series with a single annual value, usually not exceeding 100 values. To overcome this difficulty we introduce information

---

* Corresponding Author address: Email:Ismael.Baeza@uv.es.

synthetically, thereby given the characteristics of the model and the performed analyses, the polynomial interpolation is appropriate.

Then, this chapter presents a process in two stages to graduate mortality rates: the proposed graduation technique combines Piecewise Polynomial Interpolation as a previous step wavelet graduation (Baeza and Morillas, 2011). The methods validation is numerically performed and finally applied to real data of the Spanish population.

Keywords: life tables, nonparametric graduation, wavelets, polynomial interpolation

1. INTRODUCTION

Human biometrics, as part of Actuarial Statistics, works with the mortality data from a population. It uses the mortality tables to study survival. The mortality tables collect information on the variable age of death, the population at risk of dying, the number of persons who survey or die at any age or the probability of dying between ages.

This work focuses in the study of the risk of death at each age x. It is relevant to remark that the value of the risk of death it is unknown (generally). It is important to emphasize (for the demographic and actuarial fields) the importance that has to obtain good estimates of the true probabilities of death. The estimation of the subjacent risk can be used to estimate premiums of several types of insurance policies; or to estimate the mathematics reserves (obligatory) that an insurance company must set aside; or to estimate the Sustainability Factor to determine the age of the retirement, or the quantity of the pensions (in the spanish case) (Meneu-Gaya, Devesa-Carpio and Nagore-Garcia, 2013).

The study of the age of death is simplified if we consider that the variable is continuous and we assume: (i) there exists a real rate (the risk of death) but it is observed jointly with random fluctuations. Also, we only can perceive the sum of both parts, real and fluctuation;. Also, we only can perceive the sum of both parts, real and fluctuation; (ii)We assume that the rate has a structural behavior (view (Ayuso, Corrales, Guillem, Pérez-Marín and Rojo, 2007)). The first hypothesis justifies the extensive development of techniques graduation. In this sense, this work proposes a method based in the wavelet decomposition (with multiresolution schema), combining approximation by Thresholding and Piecewise Polynomial techniques.

Haberman y Renshaw (1996) define graduation as "the principles and methods by which a set of observed (or gross) probabilities are adjusted to provide a smooth base that will allow us to make inferences and also practical calculations of bonuses, reserves, etc ..."

Graduation is necessary and has an eminently statistical estimation nature. London (1985) explains the reason why we have to change and therefore graduate our initial estimates sequence obtained. This is because for each specific period, given the corresponding data, we can obtain the sequence of initial estimates ages that has sometimes abrupt changes, but it is a particular realization of the evolution of mortality. It should be satisfied that the difference between the probabilities of death of two consecutive age is not excessively high, requiring the setting of a function that meets that condition.

In literature we can find different types of techniques graduation. We can split these techniques as Parametric and non-parametric. Parametric graduation tries to find the parameters of a function that adjusts the rate. We list some examples nextly. De Moivre in

1724 and tries to model; Gompertz (1825) represents an exponential growth for mortality; Makeham (1860) adds a constant component A > 0 to this exponential growth; Weibull (1939) suggests that the mortality force grows as a power of t rather than exponentially. These laws apply only to adult ages, and many fail to represent the hump accidents in adulthood. Later, Heligman and Pollard (1980) obtained encouraging results for the full width of the interval of life. Other techniques are the semiparametric such as Forfar, McCutcheon and Wilkie (1988) where they make a brief introduction by splines graduation. The last group of techniques, non-parametric, is characterized why non functional form for the behavior of the data is assumed. In this case the mortality rates are obtained by applying smoothing methods, and similar, those combine adjacent death rates (for example kernel graduation). Some classical examples can be found in (Copas and Haberman, 1983), (Felipe, Guillem and Nielsen, 2001) and (Gavin, Haberman and Verrall, 1993). Recently, (Baeza and Morillas, 2011) propose the (new) wavelet graduation method as a non-parametric technique.

This work focuses on this latter type of techniques (non-parametric) and its objective is to generalize the results obtained in Baeza and Morillas (2011), which is useful usually for all ages above 30, or in specific mortality behaviors.

In the next section the wavelet graduation and its problems to be applied in all range of ages are briefly presented. In section 3 we introduce the Wavelet-Polynomial graduation and we compare it with kernel graduation in the evaluation of the Heligman and Pollard law. An application to real data is also reported. Finally, in section 4 there are some concluding remarks.

2. Basics and the Wavelet Graduation

This section exposes a classical way to describe and to interpret the mortality phenomenon, and its graphical representation. However, the main part of this section it is devoted to introduce some basics concepts on wavelet, and the wavelet-graduation.

2. 1. The Representation of the Mortality

As we say before, the life table summarizes a concret experience of mortality observed. The main biometric function used to analize the mortality experience and introduces it to other procces is the risk of death, q_x. An important characteristic of the observed q_x is the smoothness. To determine the smoothness level, the graphical representation is used in a first step. So, the mortality experience it is represented graphically in logarithmic scale. The experience shows that the behavior is similar in several regions or periods. The main difference is due at key points in the graphic representation. We can see Figure 1 (on a logarithmic scale) an example of mortality experience with actual data provided by the Spanish National Statistics Institute (INE) and Heligman and Pollard's theoretical law.

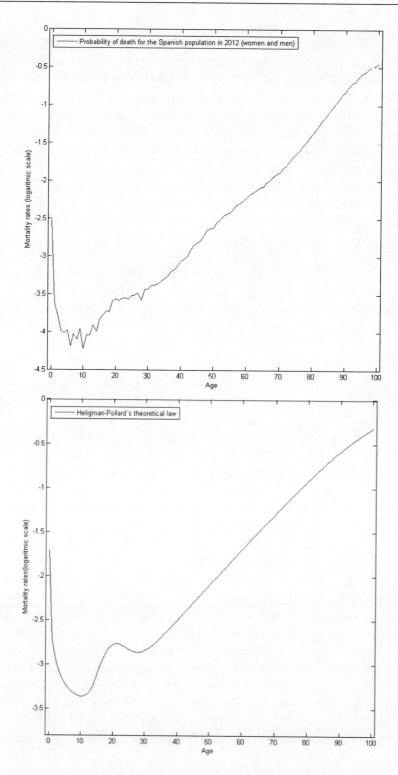

Figure 1. Probability of death for the Spanish population in 2012 (left) and Heligman and Pollard's theoretical law (right). Source: Authors using data published by INE.

Classically, we can split the experience of death in some region into three components (view Figure 1): adaptation to environment, social hump and natural mortality. The first part represents infant mortality (fast decay), the second one represents mortality in adult ages which includes death by accidents or maternity, named social or accident hump, and characterized by an excess of the mortality risk at adult ages with respect to *natural* mortality. The third component reflects the increased risk of death due to natural causes.

As we say at Introduction section, in this work it is assumed that every experience is composed of two terms (additives): the true values of the series (unknowns), and the random fluctuations. The wavelet graduation treats to recovery the true values of the biometric function considered; it is usual to treat the random fluctuations as noise to use some techniques of another fields (engineering). Let us briefly what is a wavelet.

2.2. Wavelet and Graduation

The "wavelet techniques" are a set of techniques used with several purposes. All of them using the concept of wavelet family and there exists two main approximations, the continuous and the discrete one.

First, we introduce several aspects related with the notation and some concepts. In this case we consider $L^2(\mathbb{R})$ equal to the set of all square integrable (real) functions. From a continuous point of view, a wavelet basis in $L^2(\mathbb{R})$ is a family of functions generated from dilations (or changes of scale) and translations of a generator element, $\psi(t)$, called wavelet mother. $\psi(t)$ is a function of real variable t which should range in time and well localized (decay to zero when the variable t→∞). From the mother wavelet, a family of elements can be constructed via the expression $\{\psi_{a,b}(t), a > 0, b \in \mathbb{R}\}$::

$$\psi_{a,b}(t) = \frac{1}{\sqrt{a}}\psi\left(\frac{t-b}{a}\right), \quad a,b \in \mathbb{R}, a > 0$$

The scaling parameter a is associated to the stretching or compression of the mother wavelet. Note that $\psi_{a,0}(t)$ keeps the same shape than $\psi(t)$ but in a different support. The translation parameter b "locates" temporary the distribution of energy.

The functions $\psi_{a,b}(t)$ are used to define the Continuous Wavelet Transform of $f(t)$ via the following expression:

$$W_f(a,b) := \langle f(t), \psi_{a,b}(t) \rangle = \frac{1}{\sqrt{a}} \int_{-\infty}^{\infty} f(t) \bar{\psi}\left(\frac{t-b}{a}\right) dt$$

In the same way we can define the Inverse Continuous Wavelet Transform, $\widehat{W_f}$, that verifies $\widehat{W_f} W_f = f$.

The most common way to make these dilations is through dyadic values, that is, taking $a = 2^j$ and then, $b = 2^j n\}$. For each b (depending of j and n), W_j denote the closure in

$L^2(\mathbb{R})$ of the linear span $\left\{\psi_{j,n}(t) = \frac{1}{\sqrt{2^j}}\psi\left(\frac{t-2^j n}{2^j}\right)\right\}_{(j,n)\in\mathbb{Z}^2}$, denoted it as $W_j :=$ $clos_{L^2(\mathbb{R})}\{\psi_{j,k}, k \in \mathbb{Z}\}$.

Clearly $L^2(\mathbb{R})$ can be decomposed as a direct sum of the spaces W_j. Also, for each $j \in \mathbb{Z}$, we define the closed subspaces V_j as

$$V_j = \cdots + W_{j-2} + W_{j-1}, \quad j \in \mathbb{Z}$$

These subspaces have some interesting properties. We highlight:

1. $\cdots \subset V_{-1} \subset V_0 \subset V_1 \subset \cdots$
2. $clos_{L^2}\left(\bigcup_{j\in\mathbb{Z}} V_j\right) = L^2(\mathbb{R})$
3. $\left(\bigcap_{j\in\mathbb{Z}} V_j\right) = \{0\}$
4. $V_{j+1} = V_j + W_j, \quad j \in \mathbb{Z}$
5. $f(t) \in V_j \Leftrightarrow f(2t) \in V_{j+1}, \quad j \in \mathbb{Z}$

Also (Mallat, 1980), the theory indicate that exist a function $\phi \in L^2(\mathbb{R})$ such that: (i) $\{\phi(t-n)\}_{n\in\mathbb{Z}}$ is a basis of V_0 and, (ii) all subspaces V_j are generated from dilations and translations of:

$$\phi_{j,n}(t) := \frac{1}{\sqrt{2^j}}\phi\left(\frac{t-2^j n}{2^j}\right)$$

The function ϕ is called scaling function and it is used to obtain the trend of a function f. These 'trend' information is supplemented with the information that provides the wavelet transform, and then it provides the "details". From these facts, the scaling functions jointly to the subspaces V_j (that satisfy the previous properties) provide us the called a multiresolution analysis.

The most simple and classical example of wavelet basis is based on the Haar function. In 1910, Haar (Haar, 1910) constructs a constant piecewise function such that the dilations and translations of this function ($\psi_{j,n}$ with $a = 2^j, b = 2^j n$ and $(j,n) \in \mathbb{Z}^2$) generate a orthonormal basis of $L^2(\mathbb{R})$. The expression of these functions are, to the mother wavelet function:

$$\psi(t) = \begin{cases} \frac{1}{\sqrt{2}} & \text{if } 0 \leq t < \frac{1}{2} \\ \frac{-1}{\sqrt{2}} & \text{if } \frac{1}{2} \leq t < 1 \\ 0 & \text{otherwise} \end{cases}$$

and the expression to the associated scaling function:

$$\psi(t) = \begin{cases} 1 & f \ 0 \le t < 1 \\ \frac{1}{\sqrt{2}} \\ 0 & \text{otherwise} \end{cases}$$

In the discrete framework, focus of this paper, a piecewise constant approximation can be used. In that case, $g \in V_j$ if $g \in L^2(\mathbb{R})$ such that $g(t)$ is constant for $t \in [n2^j, (n+1)2^j[$, with $n \in \mathbb{Z}^2$.

Thus the data that we work can be considered as a vector of $\mathbb{R}^N$. So, the Haar vectors are:

$$w_1 := \left(\frac{1}{\sqrt{2}}, \frac{-1}{\sqrt{2}}, 0, \cdots, 0\right) \qquad v_1^1 := \left(\frac{1}{\sqrt{2}}, \frac{1}{\sqrt{2}}, 0, \cdots, 0\right)$$

$$w_2 := \left(0, 0, \frac{1}{\sqrt{2}}, \frac{-1}{\sqrt{2}}, 0, \cdots, 0\right) \qquad v_2^1 := \left(0, 0, \frac{1}{\sqrt{2}}, \frac{1}{\sqrt{2}}, 0, \cdots, 0\right)$$

$$\vdots \qquad \qquad \vdots$$

$$w_{N/2} := \left(0, 0, \cdots, \frac{1}{\sqrt{2}}, \frac{-1}{\sqrt{2}}\right) \qquad v_{N/2}^1 := \left(0, 0, \cdots, \frac{1}{\sqrt{2}}, \frac{1}{\sqrt{2}}\right)$$

This two set of vectors, jointly form a system of N orthonormal vectors, so is, $w_n \cdot v_m = 0$, $\forall n, m = 1, \cdots, N/2$ and $n \ne m$; and $w_n \cdot w_n = v_n \cdot v_n = 1$ for all n. This implies that the family $\{v_1, v_2, \cdots, v_{N/2}, w_1, w_2, \cdots, w_{N/2}\}$ is an orthonormal basis of vectors in $\mathbb{R}^N$.

So, we can express any vector of $\mathbb{R}^N$ as:

$$f = (f \cdot v_1)v_1 + (f \cdot v_2)v_2 + \cdots + (f \cdot v_{N/2})v_{N/2} + (f \cdot w_1)w_1 + (f \cdot w_2)w_2 + \cdots$$
$$+ (f \cdot w_{N/2})w_{N/2}$$
$$= a_1 v_1 + a_2 v_2 + \cdots + a_{N/2} v_{N/2} + d_1 w_1 + d_2 w_2 + \cdots + d_{N/2} w_{N/2}$$

So, we have that

$$f = A^1 + D^1$$

where A^1 is the orthogonal projection of f onto subspace $V^1 = \text{lin}\{v_1, v_2, \cdots, v_{N/2}\}$ and D^1 is the orthogonal projection onto $W^1 = \text{lin}\{w_1, w_2, \cdots, w_{N/2}\}$, i.e., A^1 contains the behavior average of f (or trend), and D^1 contains the details. This process is known as the first level of multiresolution analysis of data f.

Now, the second step of the multiresolution analysis repeat the descomposition process applying it again, but in this case to the scaling part. The results can be expressed as:

$$a_j^2 = \frac{a_{2j-1}^1 + a_{2j}^1}{\sqrt{2}} = \frac{\frac{f_{4j-3}+f_{4j-2}}{\sqrt{2}} + \frac{f_{4j-1}+f_{4j}}{\sqrt{2}}}{\sqrt{2}} = \frac{(f_{4j-3}+f_{4j-2})+(f_{4j-1}+f_{4j})}{2} = \frac{f_{4j-3}+f_{4j-2}+f_{4j-1}+f_{4j}}{2},$$

$$d_j^2 = \frac{a_{2j-1}^1 + a_{2j}^1}{\sqrt{2}} = \frac{\frac{f_{4j-3}+f_{4j-2}}{\sqrt{2}} + \frac{f_{4j-1}+f_{4j}}{\sqrt{2}}}{\sqrt{2}} = \frac{(f_{4j-3}+f_{4j-2})-(f_{4j-1}+f_{4j})}{2} = \frac{f_{4j-3}+f_{4j-2}-f_{4j-1}-f_{4j}}{2}.$$

We obtain the orthogonal projection of f onto the subspaces $V^2 = lin\{v_1^2, v_2^2, \cdots, v_{N/4}^2\}$, and $W^2 = lin\{w_1^2, w_2^2, \cdots, w_{N/4}^2\}$, where

$$w_1^2 := \left(\frac{1}{\sqrt{2}}, \frac{1}{\sqrt{2}}, \frac{-1}{\sqrt{2}}, \frac{-1}{\sqrt{2}}, 0, \cdots, 0\right)$$
$$w_2^2 := \left(0, 0, \frac{1}{\sqrt{2}}, \frac{1}{\sqrt{2}}, \frac{-1}{\sqrt{2}}, \frac{-1}{\sqrt{2}}, 0, \cdots, 0\right)$$
$$\vdots$$
$$w_{N/4}^2 := \left(0, 0, \cdots, \frac{1}{\sqrt{2}}, \frac{1}{\sqrt{2}}, \frac{-1}{\sqrt{2}}, \frac{-1}{\sqrt{2}}\right)$$

$$v_1^2 := \left(\frac{1}{\sqrt{2}}, \frac{1}{\sqrt{2}}, \frac{1}{\sqrt{2}}, \frac{1}{\sqrt{2}}, 0, \cdots, 0\right)$$
$$v_2^2 := \left(0, 0, \frac{1}{\sqrt{2}}, \frac{1}{\sqrt{2}}, \frac{1}{\sqrt{2}}, \frac{1}{\sqrt{2}}, 0, \cdots, 0\right)$$
$$\vdots$$
$$v_{N/4}^2 := \left(0, 0, \cdots, \frac{1}{\sqrt{2}}, \frac{1}{\sqrt{2}}, \frac{1}{\sqrt{2}}, \frac{1}{\sqrt{2}}\right)$$

This process, ability us to obtain the decomposition $A^1 = A^2 + D^2$, and so on $f = A^2 + D^2 + D^1$. This result give us second level of multiresolution analysis of data f.

Iterating the process m-times, we get the m-level of multiresolution analysis.

$$f = A^m + D^m + \cdots + D^2 + D^1$$

where A^m is the orthogonal projection of f onto $V^m = lin\{v_1^m, v_2^m, \cdots, v_{N/2^m}^m\}$, and D^m is the orthogonal projection on $W^m = lin\{w_1^m, w_2^m, \cdots, w_{N/2^m}^m\}$.

Summarizing, the result of to apply the Wavelet Transform (continuous or discrete) is formed by two functions (in the continuous case) or two discrete series (in the discrete case). The first part is known as scaling part, the second one as wavelet. As we saw, scaling parts give us a first approach that includes the trend obtained. This part 'lost' the details of the initial series, but these are contained in the wavelet part. This process can be applied iteratively in the scaling part and then, we obtain a new multiresolution level.

Often, we observe an initial data, $f = \hat{f} + r$, such that $\hat{f}$ is a value that we must to estimate, and r is a random fluctuation (error or not). It is usually assumed that, in the process of the multiresolution analysis, the wavelet part (of f) contains some details necessaries to obtain f, jointly with information about the random fluctuations. So is, a part of the differences between the original series and the obtained by scaling (details) are considered disturbances. The wavelet part and the scaling part obtained calculating the wavelet transform are orthogonal, so is, the information contained in a part is not contained in the other. So the elimination or reduction of noise (random fluctuation) is linked to the treatment of wavelet part. The aim of the wavelet graduation is to reduce or even eliminate random fluctuations via thresholding, truncating some values of wavelet part, assuming certain threshold as an element that determines whether a value is or not considered as perturbation, reducing o removing it.

Complete information above wavelet can be find in (Chui, 1992) and (Daubechies, 1992).

2.3. Wavelet Graduation Problems

The wavelet graduation may have more or less significant drawbacks according to the available information or the functional relationship of the data. In the case of life tables, in Baeza and Morillas (2011), this technique only can be applied to ranges above 30 years of age. For younger age mortality curve has a non-linear relationship that complicates the analysis by the lack of information. We consider important to highlight some aspects:

- When we apply the wavelet technique, the incorporation of symmetric information in ortder to reconstruct the tails of the series introduces noise and we obtain poor graduations.
- This problem reappears if we use a wavelet family with a big support or if we make severals scaler of the process.
- Some effect, similar to the Gibbs phenomenon, has also been detected by smoothing the central area, the accident hump.

Figure 2 shows these effects.

With the main to avoid the problems described, we incorporate information. In the case of this paper, the inter-annual mortality rates are interpolated via piecewise polynomial interpolation.

3. THE WAVELET- POLYNOMIAL GRADUATION

As we have seen, graduation wavelet has problems when we apply it to the entire range of age of the biometric function. To solve this we will introduce some additional information for inter annual data. These new data will be given by a polynomial interpolation.

To test this combined graduation technique, we build 10000 "synthetic" death experiences, which are based on a particular biometric model with a numerically generated random fluctuations. In this chapter we use Heligman and Pollard's law in generating synthetic death experiences.

The process described below is carried out as many times as different experiences you want to generate.

- We start the process using theoretical probabilities of death given q_x by the Heligman and Pollard's law for and taking into account a random number of individuals l0 = 10.000, 100.000,...
- We use that the number of deaths[1] at the age of x follows a binomial distribution: $d_x \sim Bi(l_x, q_x)$ and we generate a random number given by the distribution $Bi(l_0, q_0)$. It is the number of deaths at the age 0, $\tilde{d}_0$ and we use it for the estimation $\tilde{l}_1$.

---

[1] d_x are the number of individuals that are alive at the age x but they don't at $x + 1$. $d_x = l_{x+1} - l_x$, where l_x is the number or survivor at age x.

- Then we become to generate a random number that follows a distribution $Bi\left(\tilde{l}_1, q_1\right)$. We obtain the number of deaths at age $x = 1$, $\tilde{d}_1$ and we use it for the estimation of $\tilde{l}_2$.
- Iterating this process, we generate random number from a binomial law with parameters: the estimate number of survivors in the previous stage $(\tilde{l}_x)$ and the risk of death at the age considered (q_x), estimated by the Heligman and Pollard's law. In this way, we obtain $\tilde{d}_x$ and $\tilde{l}_{x+1}$, this later one used for the next step as input of a new random number of the distribution $Bi\left(\tilde{l}_{x+1}, q_{x+1}\right)$.
- The process ends when we obtain the last value $\tilde{d}_\omega$.

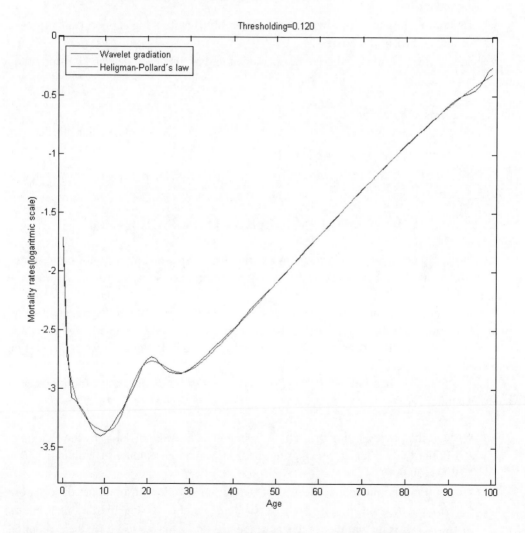

Figure 2. Gibbs phenomenon (in the social hump) and noise discontinuity at the tails.

We use the next indicator and measures to compare the Wavelet-Polynomical technique with the usual Gaussian kernel graduation:

- Mean relative indicator (MRI):

$$IRM(q) = \frac{1}{\omega+1}\sum_{x=0}^{\omega}\frac{|q_x - \hat{q}_x|}{q_x}$$

- Mean squared relative indicator (IRCM)

$$IRCM(q) = \frac{1}{\omega+1}\sum_{x=0}^{\omega}\frac{|q_x - \hat{q}_x|^2}{q_x}$$

- Whittaker-Henderson smoothness indicator (Whittaker, 1926)

$$S = |S(\hat{q}_x) - S(q_x)| \text{ where } S(q_x) = \sum_{x=0}^{\omega-2}(\Delta^2 q_x)^2$$

The first two indicators are the traditional measures for the difference of two vectors. The last indicator uses second order divided differences to measure the smoothness and it is crucial for calculating the threshold because we use the one whose reconstruction is closer to the softness given by the theoretical of Heligman and Pollard law. We can see this election in the Figure 3 (the abrupt fall of the indicator).

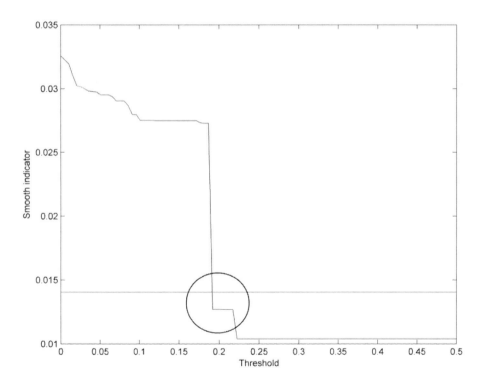

Figure 3. Evolution of the smoothness indicator of the reconstruction (by threshold) jointly with the theoretical (horizontal-red line).

In these definitions q_x denotes the theoretical probability of death that the Heligman and Pollard law provided for the preset parameters and it is used for generating the experiences of mortality; $\hat{q}_x$ denotes the graded probability, the value obtained by applying the graduation to each generated realizations. We evaluate the ability of the technique in the recovery of the true values of the function. The defined indicators suggest that the lower value is the best estimate of the theoretical probability is obtained, suggesting that another technical improvement in this regard.

Table 1 shows the results of the comparison between the Wavelet-Polynomial graduation (biorthogonal wavelet family and linear interpolation) and the Gaussian kernel graduation. In columns 3 and 4 we can see the mean value of the indicators for the 104 synthetic death experiences. The last column presents the percentage of times that Wavelet-Polynomial obtains better results than Gaussian kernel graduation. Morover, the relative differences of these indicators are higher in the cases when the kernel graduation is better than Wavelet-Polynomial technique.

Table 1. Numerical comparison Wavelet-Polynomial vs. mobile average

N	Indicator	Wavelet-Polyn.	Kernel	Better W-P
100	IRM	0,033897702	0,039032115	86,74%
	IRCM	0,003702556	0,008858977	99,84%
	S	0,001225142	0,00945352	100%
200	IRM	0,031953776	0,038258508	92.30%
	IRCM	0,002887772	0,008802846	99.98%
	S	0,001349435	0,010201342	100%
400	IRM	0,034275660	0,037640481	77,42%
	IRCM	0,003456605	0,008689914	99,93%
	S	0,002860160	0,010538267	100%
800	IRM	0,033807940	0,037275545	78,72%
	IRCM	0,003230176	0,008645842	100%
	S	0,001664736	0,010702538	100%

The table 2 shows the parameters for the Wavelet-Polynomial graduation. We opted for the biorthogonal wavelet family. Wavelet is selected by means of a criterion based on energy retention instead an exhaustive strategy like in (Baeza and Morillas, 2011). The measure is given by $H = \frac{\|\hat{q}_x\|^2}{\|q_x\|^2}$. We work with a criterion for thresholding based on (Mallat, 1980).

Table 2. Parameters used in Wavelet-Polynomial technique

N	Wavelet	Scales	Thresholding
100	Biorthogonal 3.3	2	0,15
200	Biorthogonal 3.3	3	0,2
400	Biorthogonal 3.3	4	0,25
800	Biorthogonal 3.3	5	0,3

In Figure 4 (left), are presented for the entire range of ages: the Heligman and Pollard theorical model; a random realization; and the two approximations by graduation, the kernel graduation and the Wavelet-Polynomial graduation (with N = 400 and the parameters given by Table 2). In Figure 4 (right) we observe details of these functions more closely.

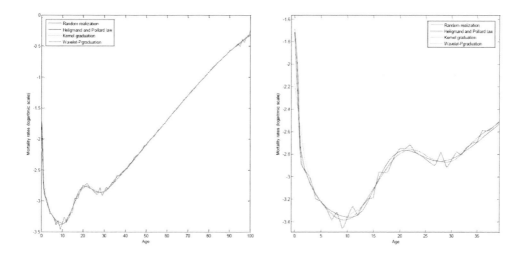

Figure 4. Left: Comparison (all range of ages). Right: Details. Source: Authors.

4. AN APPLICATION TO OBSERVED (REAL) DATA

We apply the Wavelet-Polyomial technical data to actual mortality rate of Spain to check the results of this work. As the actual rates are unknown, we cannot compare it but, using different values of Table 2 and considering the results of the previous section we believe that the approach of Figure 5 is a good reconstruction of the mortality rate. This can be used by the different agents of actuarial science in their fields.

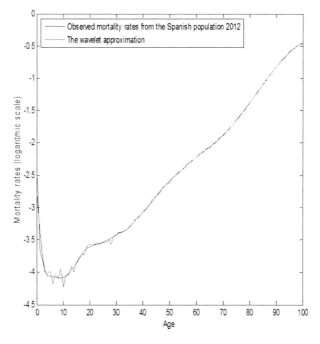

Figure 5. Spanish population 2012 mortality rates and Wavelet-Polynomial graduation (the real values are unknowns).

FINAL REMARKS

This chapter presents a process in two stages to graduate mortality rates. The process combine wavelets and polynomial interpolation trying to generalize the results obtained in (Baeza and Morillas, 2011). In the first stage the polynomial interpolation allows to incorporate additional data. The second stage use wavelets eliminate the noise (or random fluctuations) in order to reconstruct the true values of the biometric function considered. That technique can be applied to all range of ages and, in the sense of the indicators used, give us better results than kernel graduation. Also, the technique presented is more robust in the sense that, when the indicator considered is better (minus value) for the kernel graduation than Wavelet-Polynomial technique, the relative difference is higher than if we consider the reverse relation.

ACKNOWLEDGMENTS

This research was partially supported by Ministerio de Economía y Competitividad under grant MTM2012-31698.

REFERENCES

Ayuso, M., Corrales, H., Guillem, M., Pérez-Marín, A. and Rojo, J. (2007). *Estadistica Actuarial Vida*. Barcelona: UBe. [*Actuarial Statistics Life*. Barcelona: UBe.]

Baeza, I. and Morillas, F. (2011). Using wavelet to non-parametric graduation of mortalily rates. *Anales del Instituto de Actuarios, 17*, 135-164.

Chui, C. (1992). *An introduction to wavelets*. Boston: Academic Press.

Copas, J. and Haberman, S. (1983). Non parametric graduation using kernel methods. *Journal of the Institute of Actuaries, 110*, 135-156.

Daubechies, I. (1992). *Ten Lectures on Wavelets*. Philadelphia: SIAM.

Felipe, A., Guillem, M. and Nielsen, J. (2001). Longevity studies based on kernel hazard estimation. *Insurance: Mathematics and Economics, 28*, 191-204.

Forfar, D., McCutcheon, J. and Wilkie, A. (1988). On graduation by mathematical formulae. *Journal of the Institute of Actuaries, 115*, 693-694.

Gavin, J., Haberman, S. and Verrall, R. (1993). Moving weighted average graduation using kernel estimation. *Mathematics and Economics, 12*(2), 113-126.

Gompertz, B. (1825). On the nature of the function of the law of human mortality and on a new mode of determining the value of life contingencies. *Transactions of The Royal Society, 115*, 513-585.

Haar, A. (1910). Zur theorie der orthogonal en funktionensysteme. *Mathematische Annalen, 69*, 331-371.

Haberman, S. and Renshaw, A. (1996). Generalized linear models and actuarial science. *The Statistician, 4*(45), 113-126.

Heligman, S. and Pollard, J. (1980). The age pattern of mortality. *Journal of the Institute of Actuaries, 107*, 49-80.

Instituto Nacional de Estadistica. (n.d.). *Tablas de mortalidad de la población de España por año, sexo, edad y funciones[Online].* (INEbase, Ed.) Retrieved 09 02, 2015, from http://www.ine.es. [*Lifes tables for the population of Spain by year, sex, age and functions [Online].* (INEbase, Ed.) Retrieved September 2, 2015, from http://www.ine.es.]

London, D. (1985). *Graduation: The Revision of Estimates.* Coonecticut: ACTEX Publications.

Mallat, S. (1980). A theory for multiresolution signal decomposition: The wavelet representation. *IEEE Translation, 11*(7), 84-95.

Meneu-Gaya, R., Devesa-Carpio, J. and Nagore-Garcia, A. (2013). El Factor de Sostenibilidad: Diseños alternativos y valoración financiero-actuarial de sus efectos sobre los parámetros del sistema. *Economía Española y Protección Social, 63*, 63-96. [Factor Sustainability: Alternative designs and financial-actuarial its effects on system parameters assessment. *Spanish Economy and Social Protection, 63*, 63-96.]

Whittaker, E. (1923). On a new method of graduation. *Proceedings of the Edinburgh Mathematical Society, 41*, 63-75.

In: Modeling Human Behavior: Individuals and Organizations ISBN: 978-1-53610-197-3
Editors: L. Jódar Sánchez, E. de la Poza Plaza et al. © 2017 Nova Science Publishers, Inc.

Chapter 17

TRAJECTORIES SIMILARITY: A PROPOSAL AND SOME PROBLEMS

Francisco Javier Moreno[1,*], *Santiago Román Fernández*[1,†]
and Vania Bogorny[2,‡]

[1]Universidad Nacional de Colombia, Bogotá, Colombia
[2]Universidade Federal de Santa Catarina, Florianópolis - SC, Brazil

ABSTRACT

In this chapter, we focus on trajectories similarity. We present some open problems with regard to determining the similarity. For example, the similarity can be based on shape, speed, mode of transport, visited places, activities performed in the places, frequency of visits, among others. We propose a simple algorithm for determining the similarity with regard to visited places. For that purpose, we use a hierarchy of places; in this way, the analysts can determine the similarity of trajectories from different abstraction levels.

Keywords: trajectory similarity, semantic trajectories, moving objects

INTRODUCTION

The discovery of similar movement behavior from trajectory data may be useful for several domains. For example, in tourism it may help to design tourist routes. Consider, for example, some tourists that are visiting a city on a specific day. Each tourist visits a set of places $p_1, p_2, ..., p_n$. If we identify a subset of places that are visited by most tourists, we may design a route that includes all or some of these places. In this example, we are considering

---

[*] Corresponding author: Francisco Javier Moreno. Universidad Nacional de Colombia, Bogotá, Colombia. Email: fjmoreno@unal.edu.co.
[†] Santiago Román Fernández e-mail: sromanf@unal.edu.co.
[‡] Vania Bogorny e-mail: vania.bogorny@ufsc.br.

the trajectory similarity with regard to visited places (the movement of each tourist throughout the city during the day represents a trajectory). However, trajectory similarity can be considered from other points of view. For example, two trajectories may be considered similar with regard to their shape, see Figure 1.

We may also consider the similarity with regard to speed, see Figure 2. Although these two trajectories are different in shape, their average speed is the same.

Another alternative for determining trajectories similarity is to consider the mode of transport. Consider, for example, Figure 3. We can see that although the two trajectories have different shape, and their average speed is also different, they use the same means of transport. Note also that they use the same means of transport in the same order and each means of transport was used for the same amount of time (one hour). Depending on the similarity requirements, the trajectories may be considered similar, for example, if they use the same means of transport but not necessarily for the same amount of time (a threshold may be set). Frequency of use may also be a factor to be considered in trajectories similarity. Suppose, for example, that during a day, a tourist used three times a car, two times a bicycle, and four times the subway. Another tourist used four times a car, four times a bicycle, and two times a tram. The trajectories may be considered similar, for example, if the requirement is that the tourists should have in common, at least two means of transport and each of them should have been used at least two times.

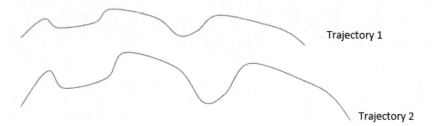

Figure 1. Two trajectories similar in shape.

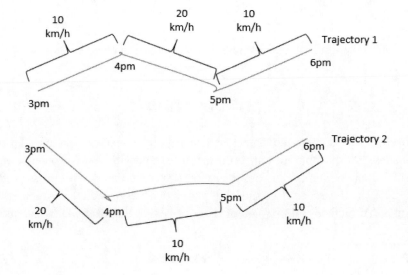

Figure 2. Two trajectories similar in speed.

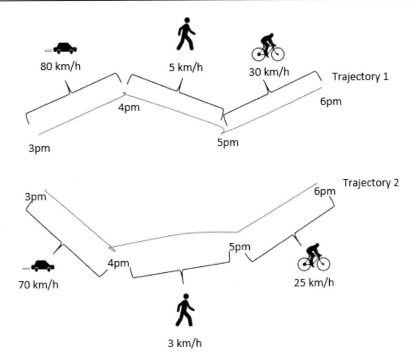

Figure 3. Two trajectories similar in mode of transport.

Informally, a trajectory of a moving object is a sequence of time-ordered points (x, y, t), where (x, y) represents the spatial coordinates and t the time. This definition corresponds to a *raw* trajectory. In the last decade, there have been several proposals that have enriched raw trajectories. For example, Spaccapietra (2008) defines a trajectory as a sequence of *stops* and *moves*. A stop represents a period of a trajectory during which an object did not move (the object was possibly visiting a place, for example, a restaurant). On the other hand, a move represents a period of a trajectory during which an object was indeed moving (the object was moving from a point A to a point B). For instance, in Figure 4, we show a trajectory with four stops (the person slept, took lunch, swam, and went to a theatre) and three moves (note that a move is defined by two stops).

We could also consider activities performed in the places. For example in Figure 5, we show two trajectories of two persons. These two persons visited the same places, but they performed different activities there. Thus, these trajectories are similar with regard to places but they are not with regard to the activities that the persons performed in these places.

In a more general way, we could say that a trajectory is a sequence of *episodes*, where the nature of each episode is defined by the analysts. For example, we could define a trajectory as a sequence of episodes, we could consider just two types of episodes *working* and *studying* and we could consider that the rest of the time the object was just moving because we are only interested in this type of episodes.

During the last few years, several approaches have been proposed to measure the similarity of raw trajectories. Among the main approaches are the DTW (Dynamic Time Warping) (Keogh and Ratanamahatana, 2004; Kruskal, 1983), developed for time series, LCSS (Longest Common Subsequence) (Vlachos, Kollios and Gunopulos, 2002), and EDR (Edit Distance on Real Sequences) (Chen, Özsu and Oria, 2005).

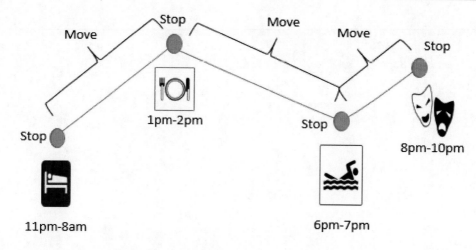

Figure 4. A trajectory represented as a sequence of stops and moves.

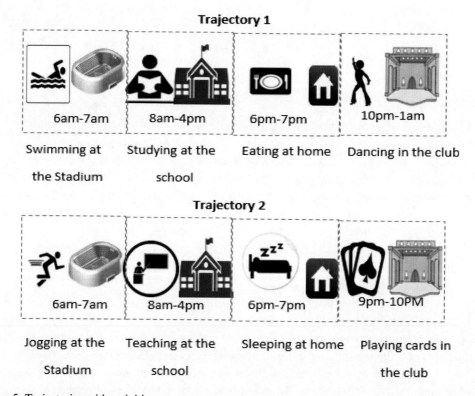

Figure 5. Trajectories with activities.

Recently, there have been other efforts to enrich raw trajectories, that is, transforming a raw trajectory into a *semantic trajectory* (Alvares et al., 2007; Parent et al., 2013). In addition to space and time, semantic trajectories have data, such as the name of the places visited by an object, and the activities performed at each place (Bogorny, Renso, de Aquino, de Lucca Siqueira and Alvares, 2014).

Several definitions can be found in the literature for *semantic trajectories* (Alvares et al., 2007) and (Bogorny et al., 2014), but for simplicity, we will consider a semantic trajectory as a sequence of important places called *stops*, as originally proposed by (Spaccapietra et al., 2008).

The first approach, proposed in 2012 by (Liu and Schneider, 2012), splits a semantic trajectory into subtrajectories and computes the semantic similarity of two trajectories based on the longest common subsequence of visited places. In their approach only a complete match is considered, that is, 1 if there is a match and 0 if there is no match on the name of the place.

Xiao, Zheng, Luo, and Xie (2012) proposed a semantic similarity measure that considers the semantics of the stops, the sequence of the visited places (stops), the travel time between the places, and the frequency that a place is visited. Two trajectories are considered to be similar if they visit the same sequence of places, several times, and with similar travel time.

In this chapter, we propose a new similarity function for semantic trajectories, where we consider a hierarchy of places to establish the similarity among trajectories. The hierarchy allows the analysts to compute similarity of trajectories in different levels of abstraction. To the best of our knowledge, our proposal is the first that considers this aspect.

TRAJECTORIES SIMILARITY

Similarly to (Lee and Chung 2011), we consider a category tree for the classification of the places (CTreeCP), where a place is a point of interest (POI). For simplicity, each place is associated with a unique category corresponding to a leaf node of the tree. In Figure 6, we show an example of a CTreeCP. The relationship between the CTreeCP nodes is hierarchical, where a child node represents a more specialized category than the category represented by its parent node.

Let P be a set of places $P = \{p_1, p_2, ..., p_m\}$ where $p_i = (pl\_id, pl\_name, pl\_category)$, pl_id is the place identifier, pl_name its name, and $pl\_category$ represents the CTreeCP category (leaf node, in our example, see Figure 6, the leaf nodes are School, University, Gym, Stadium, and Park) which is associated with the place. Thus, one place is (directly) associated with one leaf node of the CTreeCP and (indirectly) with all the ancestor nodes of that leaf node in the CTreeCP.

Example. Let $P = \{p_1, p_2, p_3, p_4, p_5, p_6\}$ be the set of places where p_1 = (1, National University, University), p_2 = (2, Body Gym, Gym), p_3 = (3, Central Park, Park), p_4 = (4, National Stadium, Stadium), p_5 = (5, Copacabana School, School), p_6 = (6, Green Park, Park). In Figure 7, we show the associations among places and the CTreeCP from Figure 6.

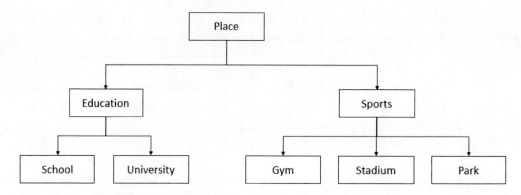

Figure 6. CTreeCP example.

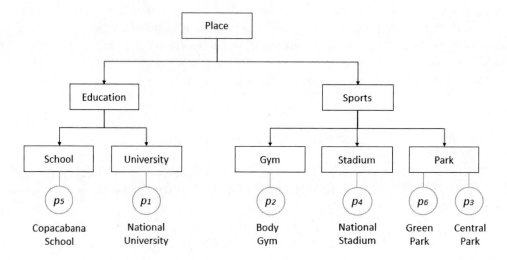

Figure 7. Associations among places and the CTreeCP.

Table 1. Events of the trajectory Tr_1

ep_i	p_i			t_i	
	pl_id	pl_name	pl_category	t_{begin}	t_{end}
				May 2 2016	
ep_1	5	Copacabana School	School	6 am	7 am
ep_2	3	Central Park	Park	10 am	12 m
ep_3	1	National University	University	1 pm	4 pm
ep_4	2	Body Gym	Gym	9 pm	11 pm

On the other hand, a trajectory Tr is a set of episodes $T_r = \{ep_1, ep_2, ..., ep_k\}$, where $ep_i = (p_i, t_i)$, $p_i \in P$ represents the place where the episode occurred, and $t_i = (t_{begin}, t_{end})$ represents the begin time and the end time of the episode, $t_{begin} < t_{end}$.

Example. Consider the trajectory $Tr_1 = \{ep_1, ep_2, ep_3, ep_4\}$ where $ep_1 = \{p_5, t_1\}$, $ep_2 = \{p_3, t_2\}$, $ep_3 = \{p_1, t_3\}$, and $ep_4 = \{p_2, t_4\}$; and where $t_1 = $ (6 am, 7 am), $t_2 = $ (10 am, 12 m), $t_3 = $ (1 pm, 4 pm), and $t_4 = $ (9 pm, 11 pm), all times correspond to May 2 2016. Table 1 illustrates the episodes of Tr_1.

To compute the similarity between two trajectories, we extend the proposal of Zhao, Han, Pan, and Yin (2009). They propose a formula to determine whether two trajectories are spatial similarity complete based on the set of POIs of each trajectory and a threshold θ.

Let POI(*nodeofInterest*, *Tr*) be the set of all places (either *directly* or *indirectly*) associated with one node *nodeofInterest* ∈ CTreeCP included in the episodes of trajectory *Tr*. The similarity between two trajectories Tr_i and Tr_j with regard to node *nodeofInterest*, Similarity (*nodeofInterest*, Tr_i, Tr_j), is calculated as follow:

- a = Cardinality(POI(*nodeofInterest*, Tr_i) intersect POI(*nodeofInterest*, Tr_j))
- b = Cardinality(POI(*nodeofInterest*, Tr_i) union POI(*nodeofInterest*, Tr_j))

Similarity(*nodeofInterest*, Tr_i, Tr_j) = a/b, $b \neq 0$. If $b = 0$ then Similarity (*nodeofInterest*, Tr_i, Tr_j) = *undefined*

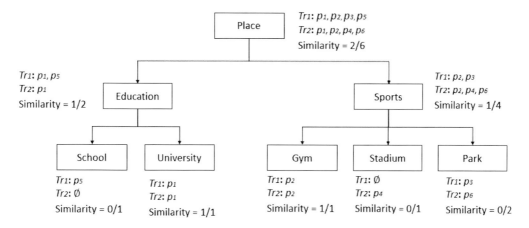

Figure 8. CTreeCP with similarity degrees for Tr_1 and Tr_2.

That is, Similarity(*nodeofInterest*, Tr_i, Tr_j) is the relationship between the total number of places common to the two trajectories associated with the node *nodeofInterest* and the total number of places of the two trajectories associated with that node. Note that Similarity(*nodeofInterest*, Tr_i, Tr_j) is *undefined* if POI(*nodeofInterest*, Tr_i) union POI(*nodeofInterest*, Tr_j) = ∅, that is, when none of the two trajectories have places associated with the node *nodeofInterest*.

Consider two trajectories Tr_i and Tr_j. We compute the similarity of each node *nodeofInterest* ∈ CTreeCP, that is, Similarity(*nodeofInterest*, Tr_i, Tr_j). In this way, the user can analyze the trajectories similarity with regard to each CTreeCP node. For instance, if *nodeofInterest* is the root of CTreeCP, then Similarity(*Place*, Tr_i, Tr_j) represents the similarity of the trajectories from a general point of view. The user can then analyze the similarity from a more specific point of view as he descends through the levels of the tree CTreeCP.

Example. Consider the trajectory $Tr_2 = \{ep_1, ep_2, ep_3, ep_4\}$ where $ep_1 = \{p_2, t_1\}$, $ep_2 = \{p_4, t_2\}$, $ep_3 = \{p_1, t_3\}$, and $ep_4 = \{p_6, t_4\}$; and where t_1 = (9 am, 10am), t_2 = (11 am, 1 pm), t_3 = (4 pm, 7 pm), and t_4 = (7 pm, 8 pm), all times correspond to May 2 2016.

Considering trajectories Tr_1 and Tr_2, the CTreeCP with the similarity of each node is shown in Figure 8. For example, the calculation of Similarity(*School*, Tr_i, Tr_j) is obtained as follow: the trajectories do not have common places with regard to this node, that is, $a =$ Cardinality(POI(*School*, Tr_1) intersect POI(*School*, Tr_2)) = 0. For computing Similarity(*Sports*, Tr_i, Tr_j) (a non-leaf node) the leaf nodes *Gym*, *Stadium* and *Park* are considered. The trajectories have one place in common (the place p_2), that is $a = 1$, and $b =$ Cardinality(POI(*Sports*, Tr_1) union POI(*Sports*, Tr_2)) = 4, thus Similarity(*Sports*, Tr_i, Tr_j) = ¼.

Next, we propose an algorithm to find the similarity between two trajectories corresponding to our method, see Listing 1.

Listing 1. Algorithm for trajectory similarity

Trajectory_Similarity(Tr1, Tr2, nodeofinterest, myCTreeCP)
Input:
Tr1, Tr2: Trajectories
myCTreeCP: CTreeCP
nodeofinterest ∈ myCTreeCP
Output:
Node nodeofinterest with computed similarity

BEGIN
//Function subtree() extracts the subtree with node nodeofinterest as root
1. mysubtree = myCTreeCP.subTree(nodeofinterest);
//Function leafNodes() extracts the set of leaf nodes of mysubtree
2. myleaves = leafNodes(mysubtree);
//Set of places of Tr1 related to nodes of interest for calculating similarity
3. places1 = ∅; *//Set of places of Tr2 related to nodes of interest for calculating similarity*
4. places2 = ∅;
5. **FOREACH** node Aux ∈ myleaves
Add to places1 the places of Tr1 associated with Aux node
Add to places2 the places of Tr2 associated with Aux node
END FOR
6. **IF** cardinality(places1) = 0 **AND** cardinality(places2) = 0 **THEN**
nodeofinterest.similarity = undefined;
ELSE
nodeofinterest.similarity = cardinality(places1 intersect places2)/ cardinality(places1 union places2);
END IF
END Trajectory_Similarity

CONCLUSION

In this chapter, we proposed a novel approach to measure the semantic similarity among trajectories of moving objects. To the best of our knowledge, our proposal is the first approach to consider different abstraction levels of the visited places, that is, a hierarchy of places. Our approach is flexible because it allows the analysts to define category trees (CTreeCP) for the classification of the places.

As future work we will consider the order of the visited places, the frequency, and the duration of the visits for computing the trajectory similarity. The order is important for trajectories that visit the same types of places, but not in the same order. For instance, consider three users U_1, U_2, and U_3, where U_1 and U_2 swim in the morning, study in the afternoon, and go shopping at night. U_3 goes shopping in the morning, swims in the afternoon, and studies at night. Although these three users perform the same activities, because of the order, trajectories of users U_1 and U_2 may be considered more similar. The frequency is interesting for similarity analysis when objects visit similar places and with similar frequency. The duration of the visits will be interesting to discover as more similar the trajectories are (they visit the same types of places). We also plan to include activities performed in the places in our similarity measure as we illustrated in Figure 5.

REFERENCES

Alvares, L. O., Bogorny, V., Kuijpers, B., de Macedo, J. A. F., Moelans, B., and Vaisman, A. (2007, 2007). *A Model for Enriching Trajectories with Semantic Geographical Information*, New York, NY, US.

Bogorny, V., Renso, C., de Aquino, A. R., de Lucca Siqueira, F., and Alvares, L. O. (2014). CONSTAnT – A Conceptual Data Model for Semantic Trajectories of Moving Objects. *Transactions in GIS*, 18(1), 66-88. doi:10.1111/tgis.12011.

Chen, L., Özsu, M. T., and Oria, V. (2005, 2005). *Robust and Fast Similarity Search for Moving Object Trajectories*, New York, NY, US.

Keogh, E., and Ratanamahatana, A. C. (2004). Exact indexing of dynamic time warping. *Knowledge and Information Systems*, 7(3), 358-386. doi:10.1007/s10115-004-0154-9.

Kruskal, J. B. (1983). An Overview of Sequence Comparison: Time Warps, String Edits, and Macromolecules. *SIAM Review*, 25(2), 201-237. Retrieved from http://www.jstor.org/stable/2030214.

Liu, H., and Schneider, M. (2012, 2012). *Similarity Measurement of Moving Object Trajectories*, New York, NY, US.

Parent, C., Spaccapietra, S., Renso, C., Andrienko, G., Andrienko, N., Bogorny, V., ... Yan, Z. (2013). Semantic Trajectories Modeling and Analysis. *ACM Comput. Surv.*, 45(4), 42:41--42:32. doi:10.1145/2501654.2501656.

Spaccapietra, S., Parent, C., Damiani, M. L., de Macedo, J. A., Porto, F., and Vangenot, C. (2008). A Conceptual View on Trajectories. *Data Knowl. Eng.*, 65(1), 126-146. doi:10.1016/j.datak.2007.10.008.

Vlachos, M., Kollios, G., and Gunopulos, D. (2002, 2002). *Discovering similar multidimensional trajectories*.

Xiao, X., Zheng, Y., Luo, Q., and Xie, X. (2012). Inferring social ties between users with human location history. *Journal of Ambient Intelligence and Humanized Computing*, 5(1), 3-19. doi:10.1007/s12652-012-0117-z.

Zhao, H., Han, Q., Pan, H., and Yin, G. (2009, 2009). *Spatio-temporal Similarity Measure for Trajectories on Road Networks*.

Chapter 18

A TENSOR MODEL FOR AUTOMATED PRODUCTION LINES BASED ON PROBABILISTIC SUB-CYCLE TIMES

E. Garcia[1,] and N. Montes[2,†]*
[1]Ford España Polígono Industrial Norte Almussafes S/N, Valencia, Spain
[2]Universidad CEU Cardenal Herrera, Alfara del Patriarca, Valencia, Spain

ABSTRACT

This chapter presents a tensor model for production lines based on stochastic sub-cycle times. In this new model, each machine that composes an automated production line is subdivided into subparts. Each machine part has a cycle time with a probabilistic distribution due to an automated machine does not make the same task exactly equal and with the same time in each repetition.

Therefore, the probabilistic cycle time of each machine is a concatenation of probabilistic cycle times where each one could be different probabilistic model distribution. Once the modelled cycle time for each machine part is obtained, a discrete finite state machine is used to model the whole production line. The simulation results give us the production rate, usually expressed in Jobs per Hour (JPH).

In order to validate the model, a welding line located in a Ford Factory at Almussafes (Valencia) is used. A welding unit is isolated and tested in order to measure sub-cycle times. The simulation results provides the maximum production rate that coincides with the ERR (Engineering running rate), that is the based-experience maximum production rate defined by the plant engineers.

At the end of the chapter, a discussion about the amount of applications that this new mathematical model could have in the industry is provided.

Keywords: manufacturing systems, machine variability, bowl phenomenon, throughput line

---

[*] E-mail: egarci75@ford.com.
[†] Universidad CEU Cardenal Herrera. C/ San Bartolomé 55, E-46115, Alfara del Patriarca, Valencia (Spain). Email: nicolas.montes@uch.ceu.es.

INTRODUCTION

A production line is an arrangement of machines or a set of sequential operations established in a factory whereby a product moves along while it is being built or produced. Each machine or worker performs a particular job that must be finished before the product moves to the next position in the line. The design of such lines is of considerable importance, (Battaïa et al. 2013). There are a large number of crucial decisions to be made in flow line design as, product design, process selection, line layout configuration, line balancing, machine selection, available technology, etc. Usually, these problems are considered one at a time because of their complexity, (Battaïa et al. 2013).

The last and crucial step in the process design is the line balancing, (Battaïa et al. 2013). Here tasks are assigned to the workstations and resources that will be employed on the line (this is a complex combinatorial problem and the solution mostly determines the efficiency of the line designed). Due to the relevance of this task, a large number of researchers have been working on this topic ((Battaïa et al. 2013) represents a state of the art of the matter). Depending on industrial environments, there are solutions to a number of product models, line layout, tasks and their attributes, workstations and their attributes, etc, see (Battaïa et al. 2013).

Currently, one of the important topics under assembly line design and balancing is the task processing time variability engendered by the following factors, (Gurevsky et. at 2012): instability of operators performing tasks with respect to work rate, skill and motivation; different material composition of product items; changes in product and workstation characteristics; as well as failure sensitivity, (Gurevsky et. at 2012). In particular, more papers were published in the last years about learning and ageing effect, see (Janiak et al. 2011), where the mean value and standard deviation is often used to model task times.

But, what happens when the line is designed and installed in the factory?. The designers defined the maximum production rate, mainly in jobs per hour (JPH), knowing as "*Engineering Running Capacity*" (ERC). The goal of the factory employees is to achieve this maximum production rate, see Figure 1, defining the throughput of the line. Reality shows that the ERC is extremely difficult or impossible to achieve so the factory employees define a new based-experience maximum production rate and is known as the "Engineering Running Rate" (ERR). The literature does not offer a reason about this throughput reduction, being the objective of the present chapter.

The present chapter develops a novel tensor model that will allow us to demonstrate that the ERR is due to the machine variability parts. For these propose, this new model introduces two new data classifications, the *mini-term* and the *micro-term*. The literature classified the data used in the analysis into *long-term* and *short-term*. The difference between both terms has been addressed in (Li et al. 2009). Long-*term* is mainly used for process planning, while *short-term* focuses primarily on process control. Therefore, following the definition in (Li et al. 2009), the *short-term* is referred to an operational period not large enough for machines' failure period to be described by a statistic distribution.

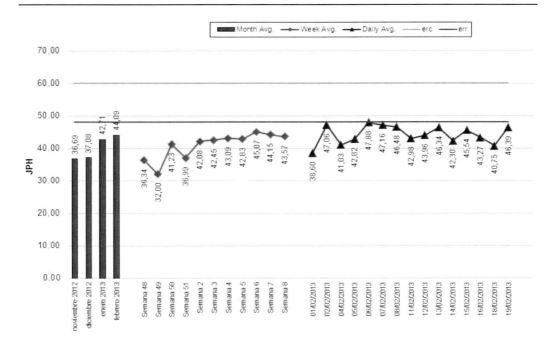

Figure 1. Jobs Per Hour produced in a real line production VS idealized production rate. Note: ECR (engineering running capacity). ERR (engineering running rate).

In order to test the proposed model, a real welding line is modelled. In particular, a real welding line in Ford S. A. located at the Almussafes factory. This line has 35 welding units distributed in 8 workstations. The simulation results provide the loss of jobs per hour due to machine variability.

The chapter is organized as follows. Section 2 presents a tensor model to compute the *long-term* and *short-term* by means of the *mini-terms* and *micro-terms*. Section 3 presents a real case study, which is a welding unit where *mini-terms* are measured experimentally. Section 4 presents a model of a real welding line and the simulation result. Section 5 presents a discussion on the results and preliminary future results. Section 6 concludes the chapter with an emphasis on future research challenges.

MATHEMATICAL MODEL. FROM MICRO-TERM TO LONG-TERM

The literature classifies the data used in the analysis into *long-term* and *short-term*. *Long-term* is mainly used for process planning, while *short-term* focuses primarily on process control. There is abundant literature for *long-term* analysis in comparison with the literature that uses *short-term* data. Therefore, following the definition in (Li et al. 2009), the *short-term* is referred to an operational period not large enough for a machines' failure period to be described by a statistic distribution. The machine's cycle time is considered *short-term*.

The present study redefines *short-term* into two new terms, *mini-term* and *micro-term*. A *mini-term* could be defined as a machine part, in a preventive maintenance policy or in a breakdown, where it could be replaced in an easier and faster manner than another machine part subdivision. Also a *mini-term* could be defined as a subdivision that allows us to understand and study the machine behaviour. In the same way, a *micro-term* is defined as each *mini-term* part that could be divided itself. In general, Factory Plan time can be represented as a tensor;

$$F = \sum_{1 \leq l \leq p} \sum_{1 \leq s \leq k} \sum_{1 \leq M \leq j} \sum_{1 \leq m \leq i} \tau_{l,s,M,m} \qquad (1)$$

where $\{\tau_{s,i,j,k}\}$ for $1 \leq l \leq p$, $1 \leq s \leq k$, $1 \leq M \leq j$, $1 \leq m \leq i$ are independent pseudorandom times with median and variance

$$\mathbb{E}[\tau_{l,s,M,m}] = \mu_{l,s,M,m} \quad \text{Var}[\tau_{l,s,M,m}] = \sigma^2_{l,s,M,m} \qquad (2)$$

and where m corresponds to the number of *micro-terms*, M corresponds to the number of *mini-terms*, s corresponds to the number of *short-terms* and l corresponds to the number of *long-terms*. In the same way a workstation with k *short-terms* (machines) working in serial can be defined as;

$$W = \sum_{1 \leq s \leq k} \sum_{1 \leq M \leq j} \sum_{1 \leq m \leq i} \tau_{s,M,m} \qquad (3)$$

where the cycle time for each *short-term* can be defined as

$$T^s_{TC} = \sum_{1 \leq M \leq j} \sum_{1 \leq m \leq i} \tau_{s,M,m} \qquad (4)$$

for $1 \leq s \leq k$, and

$$T^M_{T_{M,s}} = \sum_{1 \leq m \leq i} \tau_{s,M,m} \qquad (5)$$

for $1 \leq s \leq k$, and $1 \leq M \leq i$. We can write then

$$W = \sum_{1 \leq s \leq k} T^s_{TC} = \sum_{1 \leq s \leq k} \sum_{1 \leq M \leq j} T^m_{T_{M,s}} \qquad (6)$$

If the workstation has k *short-terms* (machines) working in parallel, the model can be rewritten as;

$$W = \text{Max}\{T^s_{TC}\} \; \forall \; s \in [1, k] \qquad (7)$$

In both cases, as the times $\tau_{s,m,\mu}$ are pseudorandom and independent, we have;

$$\mu^s_{TC} = \mathbb{E}[T^s_{TC}] = \sum_{1 \leq M \leq j} \sum_{1 \leq m \leq i} \mu_{s,M,m} \qquad (8)$$

and

$$(\sigma^s_{TC})^2 = \text{Var}[T^s_{TC}] = \sum_{1 \leq M \leq j} \sum_{1 \leq m \leq i} \sigma^2_{s,M,m} \qquad (9)$$

for $1 \leq s \leq k$, and

$$\mu^s_{T_{M,s}} = \mathbb{E}\left[T^s_{T_{M,s}}\right] = \sum_{1\leq m \leq i} \mu_{s,M,m} \qquad (10)$$

and

$$\left(\sigma^s_{T_{M,s}}\right)^2 = \mathrm{Var}\left[T^s_{T_{M,s}}\right] = \sum_{1\leq m \leq i} \sigma^2_{s,M,m} \qquad (11)$$

Then, we have that

$$\mu^s_{TC} = \mathbb{E}[T^s_{TC}] = \sum_{1\leq m \leq i} \mu^s_{T_{M,s}} \qquad (12)$$

and

$$(\sigma^s_{TC})^2 = \mathrm{Var}[T^s_{TC}] = \sum_{1\leq m \leq i} \left(\sigma^s_{T_{M,s}}\right)^2 \qquad (13)$$

for $1 \leq s \leq k$. Now, the next step is to simulate the workstation joined with the other ones. The common way is to use a simplified machine state, see Figure 3, (Leal et al. 2011).

There are three possible workstation states, "Working," "Starving" and "Blocking." If the current station is in "Working" state and the work is finished, it checks the following station, if it is in "Starving" state, the finished part of product is delivered to it and the state of the current station is free to receive another job. If the next station is in "Working" state when the current finishes its work, the current station changes its state to "Blocking," that is, blocking itself until the next station is free. If the current station is free to receive another part, it checks the previous station. If the previous station is in "Working" state, the current state changes to "Starving" state waiting itself until the previous station has a part to work on. If the previous station is in "Blocking" state, the current station receives the part and the current state changes to "Working." There is always work available for the first station and there is always the final product for extract from the last one.

When the simulation starts, every station state is set to "Starving," but the first one whose state is set to "Working" state. The simulation loop runs at predefined step time (Δt). For each step time, the cycle time of each workstation decreases until the cycle time is zero, meaning that the work is finished and the events are triggered.

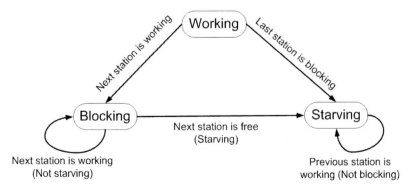

Figure 2. Simplified machine state for the workstation.

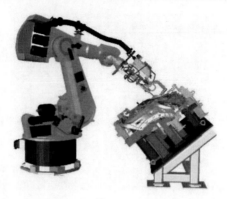

Figure 3. Welding station.

The cycle time of a workstation has a probabilistic distribution $T_{TC}^s(\mu_{TC}^s, \sigma_{TC}^s)$ that depends on *mini-terms* and *micro-terms,* see equations (8) and (9). Then, when new job is started in a workstation, a new cycle time is generated based on the probabilistic distribution. The jobs per hour produced by the line, JPH (μ_{JPH}, σ_{JPH}), are computed using long time simulation. If we increase the time simulation for the lifetime of the factory, it is possible to compute the total production.

CHARACTERIZATION OF *MINI-TERM* IN A REAL WELDING STATION

The goal of the present study is to analyse the effect of some *mini-terms* on the throughput of the line. For this propose, a car welding station is taken as an example. The welding station is one of the most relevant stations because there are 4.500 welding points in a car. The welding line is composed by welding workstations that in itself has welding stations working in parallel. A robot arm and a welding clamp compose a welding station, see Figure 3. This welding station was isolated for the welding line in order to analyse, understand and measure the results presented in this section.

The behaviour of the welding station is simple. First, the robot arm moves the welding clamp to the point to weld. Then, a pneumatic cylinder moves the welding clamp in two phases: One to approximate the clamp and the second one to weld. The pressure applied by the clamp is controlled by a control system.

Each of these devices need a certain time to develop their task and within each of these devices, there are also components that also need a certain time to develop their own tasks. In order to analyse it, the welding unit is divided in three *mini-terms*, the robot arm, the welding clamp motion and the welding task.

Table 1. Experimental test measurements for each mini-term and the total cycle time. 6 repetitions

	Mini-term Robot arm (μ, σ)	Mini-term Clamp (μ, σ)	
		Motion	Welding
Six cycle times	(35.55, 0.67)	(2.49, 1.15)	(8.62, 2.05)
Single cycle times	(1, 0.11)	(0.42, 0.47)	(1.44, 0.84)

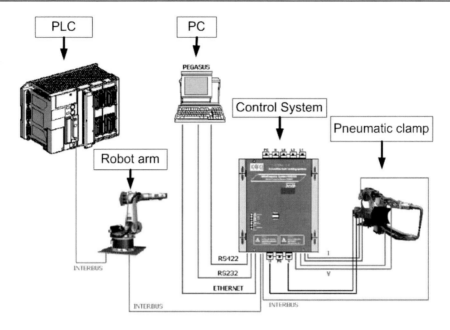

Figure 4. Experimental setup for the welding station.

Figure 4 shows the experimental setup to measure the cycle time of each *mini-term* in the welding station where the PLC and the PC are used to measure the time.

The experimental test is quite simple. The robot arm, starting from a predefined initial point, moves the clamp to a predefined welding point; the clamp is closed and develops the welding task. Due to the welding motion and the welding task that are low time consuming, the task is repeated 6 times.

The experimental methodology is as follows. The clamping task is to weld the same point 6 times in order to obtain enough time precision. The robot arm trajectory is the same in all the movements. Then, the clamping task is repeated 40 times in order to obtain a sufficient number of samples to measure the mean value and the standard deviation for each *mini-term*. Table 1 shows the experimental result measurements for each mini-term. (μ, σ).

Table 1 shows also the computed single cycle time by means of the next equations;

$$\mu = \frac{\mu}{6} \quad \sigma = \sqrt{\frac{\sigma^2}{6}} \tag{14}$$

In the case of a robot arm, the base time of the robot arm cycle time is the normal cycle time, giving us the cycle time per second, that is;

$$\mu = \frac{\mu}{\mu_{opt}} \quad \sigma = \sqrt{\frac{\sigma^2}{\mu_{opt}}} \tag{15}$$

where μ_{opt} is the total robot arm motion.

A REAL WELDING LINE. MODELLING AND SIMULATION

Previous section shows how the welding unit and, machines in general, have a probabilistic cycle time behaviour. Now, in the present section we analyse how many jobs are lost in a real line due to these pathologies. For this propose, a real welding line in Ford S.A. located at the Almussafes factory is selected, see Figure 5.

In a real welding line like this, there are welding workstations where, each one has welding stations working in parallel and sometimes in serial. Each welding station makes some welding points in the same cycle time. It is possible to find 1, 2, 4 or at least 6 welding station in the same workstation, where each one makes up to 19 welding points, see Table 3. In our particular case, our welding line has 8 workstations where 3 are for 6 welding units, 4 with 4 welding units, and 1 for 1 welding unit, see Figure 6.

The welding line was installed in 1980. The staff group that designed the line defined the maximum running capacity, ECR (engineering running capacity), 60 JPH. However, the plant engineers have another maximum running capacity, that is the ERR (engineering running rate), in this case defined in 51 JPH. And our daily production to reach is GRR (Get Ready Requirement) our (28,9 JPH). The GRR means market requirements, i.e., customers' orders. The Next figure shows the production rate of the welding line.

Table 2 shows the modelling of the welding line, where the rows show the *mini-terms* for each robot in each station, robot motion, welding motion and welding task respectively, as well as the offset. Robot motion means how many seconds the robot is moved, the welding motion and welding task means how many welding points the welding clamp has to perform. The offset means how many seconds the welding station must wait for another station to do the job.

In order to simulate the welding line, a state machine simulator is developed. There are three possible workstation states "Working," "Starving" and "Blocking," see Figure 2 and Figure 8. The loop is updated with an incremental time of 0.01 seconds. In the simulated welding line, there is always a job in the first workstation, so that the blocking state is not possible in the first station. In addition, all the jobs finished in the last workstation are retired, so that the "Starving" state in the last workstation is not possible. The loop starts with all the stations in the "Blocking" state.

Figure 5. Welding line analysed in the present section. It is located in a Ford Factory in Almussafes (Valencia).

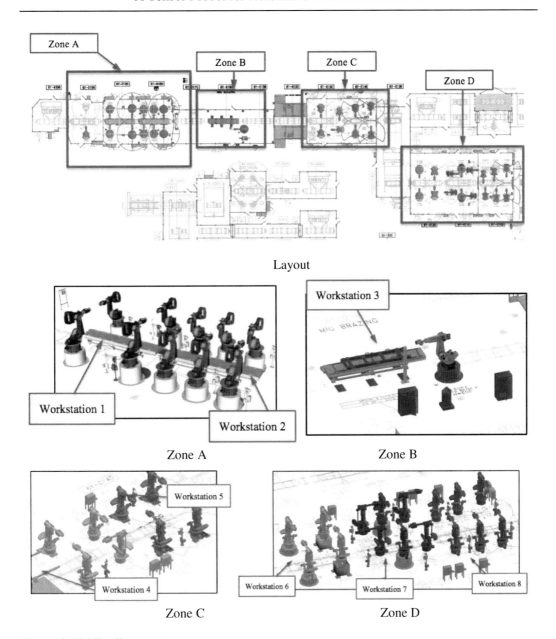

Figure 6. Welding line.

The cycle time of each welding unit is computed introducing the repetitions in Eq (6), that is;

$$W = \sum_{1 \leq s \leq k} T_{TC}^s = \left(\sum_{1 \leq s \leq k} \left(\sum_{1 \leq M \leq j} r_M \cdot T_{T_{M,s}}^M \right) + \xi \right) + \varphi \qquad (16)$$

where r_M are the repetitions of each *mini-term m*, ξ is the offset and φ is the transport time. This simulation is executed during 50 hours and the mean and variance of the jobs produced in each hour is computed. The result is, (51, 1.05) JPH. The mean value is equal than the ERR, as we wanted to prove.

Table 2. Modelled line station

		Wu1	Wu2	Wu3	Wu4	Wu5	Wu6
Workstation 1	Robot motion (Sec)	14	13.26	20.84	22.84	16.52	13.72
	Welding motion (Units)	9	9	6	6	18	18
	Welding task (Units)	9	9	6	6	18	18
	Offset(Sec)	0	11	11	0	0	0
Workstation 2	Robot motion (Sec)	24.76	21.26	---	---	9.52	11.52
	Welding motion (Units)	9	9	---	---	18	18
	Welding task (Units)	9	9	---	---	18	18
	Offset (Sec)	0	4	---	---	0	0
Workstation 3	Robot motion (Sec)	---	---	19.56	---	---	---
	Welding motion (Units)	---	---	9	---	---	---
	Welding task (Units)	---	---	9	---	---	---
	Offset (Sec)	---	---	0	---	---	---
Workstation 4	Robot motion (Sec)	13.96	16.54	16.54	17.4	---	---
	Welding motion (Units)	14	11	11	10	---	---
	Welding task (Units)	14	11	11	10	---	---
	Offset (Sec)	0	0	0	6	---	---
Workstation 5	Robot motion (Sec)	16.26	15.12	20.68	21.4	---	---
	Welding motion (Units)	9	8	12	10	---	---
	Welding task (Units)	9	8	12	10	---	---
	Offset (Sec)	0	2	0	0	---	---
Workstation 6	Robot motion (Sec)	20.26	11.02	14.12	11.34	---	---
	Welding motion (Units)	9	18	8	6	---	---
	Welding task (Units)	9	18	8	6	---	---
	Offset (Sec)	0	0	0	0	---	---
Workstation 7	Robot motion (Sec)	11.66	10.10	12.38	8.52	7.52	19.26
	Welding motion (Units)	19	15	17	18	18	9
	Welding task (Units)	19	15	17	18	18	9
	Offset (Sec)	0	0	0	0	0	0
Workstation 8	Robot motion (Sec)	12.24	13.10	12.38	14.10	13.96	15.68
	Welding motion (Units)	16	15	17	15	14	12
	Welding task (Units)	16	15	17	15	14	12
	Offset (Sec)	0	0	0	0	0	0

DISCUSSION AND PRELIMINARY FUTURE RESULTS

The simulation result demonstrates that the reason for the loss of jobs is the machine variability, and ERC cannot be achieved due to the *mini-term* and *micro-term* time deviation of each machine. The simulation also allows us to test how sensitive the throughput of the line is due to *mini-term* variability. For instance, if we replace the deviation of each *mini-term* for -0.01 sec, the simulation gives (57, 0.47) JPH, 4 Jobs below the ERC and 6 more of the ERR.

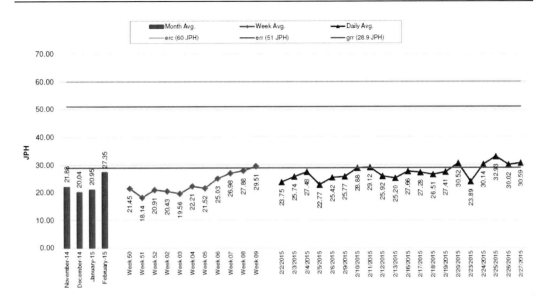

Figure 7. Jobs Per Hour produced in a real welding line VS idealized production rate. Note: ECR (engineering running capacity). ERR (engineering running rate). GRR (get ready requirement).

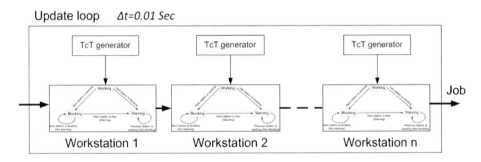

Figure 8. State machine workstation simulation. Following Eq (7), the cycle time for each workstation is the maximum cycle time of each welding station that works in parallel, indicating the slower welding unit and the bottleneck for a particular workstation. The transfer time is added to the cycle time. This time is the time required to transport the car from one workstation to the next one (12 seconds).

The mathematical model and the results presented in this chapter could be a deep impact in the production industry. Nowadays, the maintenance policy is to replace a machine part when it is broken, enough damaged for the worker detection or when the lifespan determined by the manufacturer is nearest to finish. However, there are an ignorance about what happen about the sub cycle-time and how many jobs are loosed since the machine part starts to work to the moment before to break.

By means of the same experimental setup shows in Figure 4, we test one of the most common failures in the machine production line, that is the loose of pressure in a pneumatic circuit. The pressure drop causes a delay or malfunction in the pneumatic devices to be operated. This pathology could be produced by many facts such as a simple pore that produces a failure in the compressor. Maintenance workers detect this failure when the low pressure alarm is triggered. The experimental measurement is for the *mini-term* motion and give us (1.56, 3.74) Sec. If we introduce this data in the simulation, maintaining other *mini-*

terms without pathology, we obtain that the JPH are in the range of (40, 3.25) and (34, 7.01), depending if the failure appears in the workstation 1, welding unit 4 or in the workstation 7, welding unit 4 respectively.

CONCLUSION

This chapter presents a new tensorized model that allows us to determine how the variability of a machine part can affect the production rate in a line. The new tensor model is based on two new added terms, *mini-term* and *micro-term*. A *mini-term* subdivision is selected by the user for some reason, because a machine part needs to be replaced by maintenance workers or simply because it allows the machine to be analysed more adequately. A *micro-term* is a component that composes a *mini-term* and it can be as small as the user wishes.

The present chapter focuses its attention in a welding line located in a Ford Factory at Almussafes (Valencia). This welding line has an ERC of 60 JPH. However, the employers redefines their own maximum capacity to 51 JPH, the ERR. This variation, as we demonstrate is the present chapter, is due to machine variability, and this is the reason that ERC cannot be achieved.

The model presented in this study could be a deep impact in the production industry for some reason. It could help to analyze deterioration pathologies and their effect in the throughput. In this sense, a deep characterization of the machine part deterioration is required. Although, manufactures have a threshold for the lifespan of the parts, however, for maintenance workers, the evolution of the deterioration during the lifespan time could be crucial in the throughput of the line, as well as for the detection of pathologies with great variability, evidence is given on how the variability of machine parts can affect the throughput of an assembly line. It is important to state that the preliminary result only takes into account one pathology in one welding unit and the others are without pathologies. Most probably, in a real welding line, all the welding units have a percentage of deterioration.

ACKNOWLEDGMENTS

Authors wish to thank Ford España S. A. and, in particular, Almussafes Factory for their support in the present research.

REFERENCES

Battaïa, O., Dolgui, A. A taxonomy of line balancing problems and their solution approaches. *International Journal of Production Economics;* 142, pp. 259-277 (2013).

Gurevsky, E., Battaïa, O., Dolgui, A. Balancing of a simple assembly line under variations of task processing times. *Annals of operations research.* Vol201, Is 1 pp 265-286 (2012).

Janiak, A., Krysiak, T., Trela, R. Scheduling problems with learning and ageing effects: A survey. *Decission making in manufacturing and services.* Vol 5 pp 19-36 (2011).

Leal, F. Leal, Costa, R. F. S., Montevechi, J. A. B., Almeida, D. A., Marins, F. A. S. Marins. A practical guide for operational validation of discrete simulation models. *Pesquisa Operacional* Vol 31(1) pp 57-77 (2011).

Li, L., Chang, Q., Ni, J. Real time production improvement through bottleneck control. *International Journal of production research*; 47:21, pp 6145-6158 (2009).

Lopez, C. E., Unbalanced workload allocation in large assembly lines. 2014.

In: Modeling Human Behavior: Individuals and Organizations   ISBN: 978-1-53610-197-3
Editors: L. Jódar Sánchez, E. de la Poza Plaza et al.   © 2017 Nova Science Publishers, Inc.

Chapter 19

BUILDING LIFETIME HETEROSEXUAL PARTNER NETWORKS

L. Acedo[1,*], *R. Martí*[2,†], *F. Palmi*[1,‡], *V. Sanchez-Alonso*[3,§],
F. J. Santonja[2,¶], *Rafael-J. Villanueva*[2,∥] *and J. Villanueva-Oller*[3,**]

[1]Instituto Universitario de Matemática Multidisciplinar,
Universitat Politècnica de València, Spain
[2]Departamento de Estadística e Investigación Operativa,
Universidad de Valencia, Spain,
[3]Centro de Estudios Superiores Felipe II,
Aranjuez, Madrid, Spain

Abstract

The structure and properties of sexual-contacts networks in human populations is a topic of key interest in connection with the spread of sexually transmitted infections (STI). However, this problem has received scarce attention, and the modelling of STI epidemiology is usually based on theoretical proposals in terms of the network structure usually unvalidated, and most of the times fails to reflect the real situation. The goal of this chapter is to provide a method to build a feasible network structure from real data available in the Health and Sexual Habits Survey in Spain. In particular, we focus on solving a hard optimization problem that appears in the process of creating the links between individuals, to satisfy constraints imposed by the distribution of the number of partners for both males and females. We show that such a network can be obtained by a matching method of the bipartite graph of males and females, which takes into account a preassigned degree of connectivity. In order to perform the pairing, we apply the principle of psychological similarity, by considering that people with a certain number of partners, tend to form relationships with people with a similar

---

[*]E-mail address: luiacrod@imm.upv.es (Corresponding author).
[†]E-mail address: rafael.marti@uv.es
[‡]E-mail address: pacopalmiperales@gmail.com
[§]E-mail address: vicsana3@doctor.upv.es
[¶]E-mail address: francisco.santonja@uv.es
[∥]E-mail address: rjvillan@imm.upv.es
[**]E-mail address: jvillanueva@pdi.upm.es

number of partners. This quantity is measured by a distance function and the resulting network is built using this assortative mixing. The proposed method is applied to infer the structure of networks with 1,000,000 people or more, which is larger than any other one analyzed in previous field studies.

PACS: 87.10.Rt, 87.23.Ge

Keywords: sexual network, sexually transmitted diseases, bipartite graphs

AMS Subject Classification: 94C15, 91D30s

1. Introduction

Sexually transmitted diseases have been a major public health threat for a long time in human history. Modern concerns about STI began with the pandemic of syphilis, which spread over Europe in the early sixteenth century. Nowadays, syphilis still affects twelve million people all around the world every year, causing for example 113,000 deaths in 2010 [1]. Gonorrhea spreads at a rate of AIDS, 88 million cases each year [2], while human papillomavirus is thought to be the direct cause of 561,200 new cervical cancer cases only in 2002 [3]. The global pandemic of caused by the lentivirus HIV is perhaps the most acute and widespread in human history since it has already caused 36 million deaths worldwide and it has a pool of 35.3 million people infected by HIV in 2012.

This kind of diseases is more likely to produce large-scale pandemics than other transmissible diseases, such as respiratory or other, because the efficacy of sexual contacts for the infection is large and the infectious agent has long latency periods as in the case of HIV. Moreover, nor the carrier neither his/her partner is not aware of their exposure to it. For example, it has been estimated that around 40-50 % of contacts are capable of transmitting HPV [5]. On the other hand, some STIs are caused by oncoviruses such as Hepatitis B or HPV, which increase the death rate of people who develop the disease.

In order to understand the evolution of these diseases we need a reliable model of the underlying social network in which the pandemic builds up. Individuals who change partners or have several partners simultaneously, are the hubs favouring the spread of STI. The distribution of degrees of the nodes in the network and the average chemical path from an infected individual to a susceptible one, are important parameters controlling the final extension of a new STI in a population and the speed of its spread. However, most models are based on some assumptions, which could not be valid for certain populations. Some studies claimed that the web of human sexual contacts is a scale-free network characterized by a power-law distribution for the number of individuals with a certain degree of connectivity, k: $P(k) \propto 1/k^\alpha$ with a value of α in the range $2 < \alpha < 3$, and slightly smaller for males than for females [6]. Although, $P(k)$ provides some valuable information about the network, it is not a sufficient prescription on how to build it for a given population size. Moreover, a power-law distribution of contacts could not be representative of some populations, or could vary from country to country. For example, in the Jefferson High School's network it has been found that a densely connected core appears without the need of a high connectivity degree [7].

Some field studies have ascertained the structure of moderate size real networks of sexual contacts: In 2004, Bearman et al. published the results for a set of 800 adolescents in a midsized town of the United States [7]. They showed that the structure of this network is a big cluster with a ring and extended filaments contained most of the adolescents implying that, potentially, the infection of an individual could propagate to the whole of the population, given sufficient time and infectivity. A similar study was performed in 2007 at the Likoma Island in Malawi with the idea of predicting and explaining the expansion of HIV in sub-saharan populations [8]. That study disclosed that the sexual network contained many cycles, in contrast with the single cycle at Jefferson High School. For that reason, it was speculated that superimposed cycles could be the explanation of the high prevalence of HIV infection in small populations of Africa.

Some recent studies reveal that the evolution of partnerships is also an important factor in the transmission of STIs. In particular, they pointed that the following items should be considered: (i) the cumulative distribution of the lifetime number of partners, (ii) the distribution of partnership durations, (iii) the distribution of gap lengths between partnerships, and (iv) the number of recent partners. A method for building up networks considering these items has been developed by Schmid and Kretzschmar [9]. However, this information is not available in most surveys, and we therefore face the problem of developing reasonable models for STIs in many countries where information about sexual behaviour is scarce. For example, in the case of Spain, only data about the number of sexual contacts in a lifetime is currently known from surveys. This is sufficient for building a sexual network for the transmission of HPV or other diseases with lifetime consequences and progression. In these cases, the important fact is whether the individual have had a contact with risk of infection. The remaining aspects of the network such as the duration of partnership and the time intervals among them can be incorporated effectively into a probability of transmission parameter, α.

In this paper we propose a method to build a network of sexual contacts derived from real statistical data for the distribution of the number of partners without any assumption about its global structure. For the time being, only small networks of sexual contacts have been ascertained in detail from real data and, consequently, our method fills the gap between the small communities of individuals and the large scale. Validation of the resulting structure is a problem that cannot be directly solved but our method provides a useful tool to simulate sexually transmitted diseases and obtain some indirect evidence of its validity.

Our extensive experimentation includes examples with different networks, and it is available at http://lsp.imm.upv.es.

2. Lifetime Sexual Partner Networks

The Community of Valencia is an autonomous region of Spain. It is located in the central and south-eastern side of the Iberian Peninsula. The region is divided into three provinces: Alicante, Castellón and Valencia with a total population of 5.1 million inhabitants.

The population of the Community of Valencia can be distributed per age and sex between 14 and 65 years old as shown in Table 1 [10].

From the data in this table, it is easy to obtain that 50.71% of Valencian people between 14-65 years old are male and the remainder are females. 25.22% of males are aged 14-29, 25.23% are aged 30-39 and 49.55% are aged 40-65. In a similar way, 24.87% of females are aged 14-29, 23.99% are aged 30-39, and $51,14\%$ are aged 40-65. Following this proce-

Table 1. Population of men and women in the Community of Valencia in 2013 [10] between 14 and 65 years old

Age	Men	Women	Age	Men	Women	Age	Men	Women
14	24 365	22 975	32	44 469	41 488	49	36 946	36 820
15	24 371	22 963	33	47 120	43 330	50	36 090	35 955
16	24 340	22 892	34	48 050	43 947	51	36 114	36 437
17	24 369	22 958	35	48 146	44 178	52	35 372	35 676
18	25 616	23 810	36	47 716	43 424	53	34 198	34 592
19	26 309	24 938	37	47 141	43 833	54	33 059	33 788
20	26 352	25 293	38	45 630	41 790	55	30 239	31 490
21	26 996	25 715	39	45 278	41 388	56	29 918	31 193
22	27 693	26 670	40	44 728	41 124	57	28 103	29 705
23	28 626	27 744	41	44 209	40 863	58	28 037	29 759
24	29 527	28 470	42	43 300	40 677	59	28 914	30 337
25	30 396	29 909	43	42 710	40 487	60	26 696	28 589
26	31 647	31 079	44	42 733	40 851	61	26 166	28 378
27	33 369	32 450	45	41 128	39 155	62	26 740	29 536
28	34 501	33 485	46	40 485	38 794	63	28 574	30 671
29	37 610	35 877	47	40 638	39 765	64	26 391	28 956
30	40 216	37 759	48	38 937	37 357	65	25 508	27 707
31	42 597	40 358						

dure, we can obtain the percentage of males and females per age inside their corresponding groups. This will be useful to distribute the nodes of a network demographically. In the following, we will use this data to assign ages to the nodes in the network according to the population histogram in Valencia.

Lifetime sexual partners for an individual (LSP) is obtained from the Health and Sexual Habits Survey 2003 [11], as listed in Table 2 in terms of the age-group. Notice that we do not have more recent information on this topic.

Table 2. Proportion of males and females per number of LSP per age group

			MALES			
Age	0 LSP	1 LSP	2 LSP	$3-4$ LSP	$5-9$ LSP	10 or more LSP
$14-29$	0.107	0.207	0.131	0.225	0.168	0.162
$30-39$	0.027	0.225	0.128	0.21	0.17	0.24
$40-65$	0.019	0.268	0.14	0.193	0.163	0.217

			FEMALES			
Age	0 LSP	1 LSP	2 LSP	$3-4$ LSP	$5-9$ LSP	10 or more LSP
$14-29$	0.138	0.43	0.186	0.158	0.056	0.032
$30-39$	0.029	0.501	0.168	0.177	0.077	0.048
$40-65$	0.017	0.652	0.138	0.118	0.039	0.036

We now summarize some general features of the distribution of contacts: (i) The percentage of males and females with no partners is very similar in each age-group (ii) The women proportion with only a partner is, approximately, two times larger than men with only one partner (iii) The percentage of men with two or more partners is always larger than that of women except for women in the age-groups $14-29$, and $30-39$ in the case of two partners. The asymmetry in the behaviour of males and females should be taken into account in the construction of the network.

2.1. Network Model

In this section we propose a method to create the LSP network. It basically consists of two steps. In the first one, the nodes and their asociated degree are generated. In the second one, the edges representing the partnership are added to the network.

We consider a population of N individuals, M males and F females, $N = M + F$. Let LSP_i be the number of lifetime sexual partners for node i. The challenge in building the sexual network is to find an efficient method to assign the edges to the graph in such a way that those incident to the male nodes match those incident with the edges attached to the female nodes.

In mathematical terms:

$$\sum_{i=1}^{M} LSP_i = \sum_{j=1}^{F} LSP_j .\qquad(1)$$

Some authors say that in USA, the median number of female LSP between 15-44 years old in the period 2006-2008 is 3.2 [12, 13]. Analogously, the median number of male LSP between 15-44 years old in the period 2006-2008 is 5.1. Ideally, if we multiply the number of males between 15-44 years old by 5.1, the result should be approximately the number of females between 15-44 years old by 3.2. Taking into account that males and females are around 50% of the total population, the ideal situation becomes difficult to match. Sociologists say that males tend to overestimate the number of their sexual partners and females tend to underestimate it. Therefore, if we want to build a network where the number of male and female LSP coincide, we will have to make a decision about the average number of male LSP and to assign LSP to females in such a way that the sum of LSP of males and females coincide, and vice versa. Thus, we decide to consider the average LSP male value, $\langle k \rangle_m$, and build the network from it.

After some tests, in order to have results consistent with data of Table 2, that is, men and women with 10 or more LSP have, in fact, 10 or more LSP, the average number of male LSP with age between 14-65 in the Community of Valencia should be 4.5, at least. This is a computational condition we found in the implementation of the method in order to fulfill Eq. (1).

2.2. Semi-Random Construction

The data in Table 2 are percentages and the ones in Table 1 can be transformed into percentages dividing by the total population. Transforming these percentages in accumulative percentages by successive addition, we can determine randomly the sex, age and number of LSP of the network nodes.

From Table 2 (proportion of male LSP of age 14-29) we have the following list: $(0.107, 0.314, 0.445, 0.67, 0.838, 1)$ for the accumulated proportion of males with less or equal than a given LSP number. Now, we randomly generate a number r between 0 and

1 and assign the number of contacts, to every male node in the age group 14-29, in the network as follows:

- $r \leq 0.107$ we will say that the corresponding male does not have LSP,

- $0.107 < r \leq 0.314$ we will say that the corresponding male has one LSP,

- $0.314 < r \leq 0.445$ we will say that the corresponding male has two LSP,

- $0.445 < r \leq 0.67$ we will say that the corresponding male has three or four LSP, uniformly distributed.

- $0.67 < r \ll 0.838$ we will say that the corresponding male has five to nine four LSP, uniformly distributed.

- $0.838 < r \leq 1$ we will say that the corresponding male has 10 or more LSP.

Every node in the network is labelled by its sex and age randomly assigned according to the population histogram in Table 1. The assignment of the number of bonds, as another label of the node, is not so straightforward because we must guarantee that the condition in Eq. 1 is verified. In order to fulfill this condition we take advantage of the uncertainty of statistics reports concerning individuals with 10 or more lifetime sexual partners. Starting with the males, we assign the number of LSP up to nine partners and for ten or more proceed as follows: Let $i_{\max}$ be the number of males with 9 or less partners. The unassigned males should be $M - i_{\max}$ and the number of bonds that should be distributed among them is $M \langle k \rangle_m - \sum_{i=1}^{i_{\max}} LSP_i$. By euclidian division, this quantity can be expressed as $(M - i_{\max}) n_m + r_m$, where $n_m \geq 10$. In our procedure we assign a random number of bonds, uniformly distributed, in the interval $[10, 2n_m - 10]$ to every male with 10 or more LSP, i. e., to the $M - i_{\max}$ unassigned males.

Now, we denote as p_m the total number of bonds of the male population. We must take into account, as expressed in Eq. 1, that the total number of bonds of the female population should be the same. To impose that condition we proceed as follows: (i) Assign the number of bonds to the females with 9 or less partners following the statistical data in Table 2. (ii) The sum of all female LSP in this group of $j_{\max}$ members will be denoted by s_f. Then $n_f = F - j_{\max}$ is the number of females with 10 or more LSP. (iii) The number of bonds starting in the males and still unassigned to a female is $p_m - s_f = q_f n_f + r_f$, where $0 \leq r_f < n_f$ and $n_f \geq 10$. (iv) We assign $q_f + 1$ bonds to r_f females still unassigned and q_f bonds to the rest of $n_f - r_f$ females.

The steps of this algorithm are also enumerated in the flow diagram in Figure 1.

Notice that this procedure implies that the condition in Eq. (1) is verified. After this procedure we have obtained the following lists:

- $AgeMale[i]$ is the age of the i-th male, $i = 1, \ldots, M$.

- $AgeFemale[i]$ is the age of the i-th female, $i = 1, \ldots, F$.

- $kMale[i]$ is the number of lifetime sexual partners for the i-th male, $i = 1, \ldots, M$.

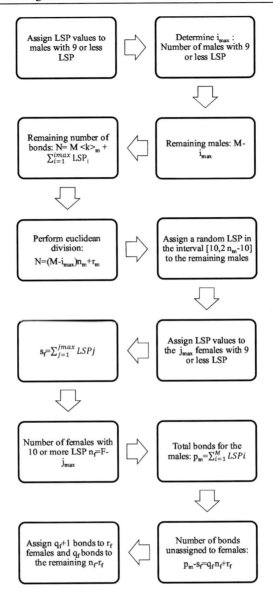

Figure 1. Flow diagram for the algorithm corresponding to the assignment of a number of LSP to every male and female in the network.

- $kFemale\,[i]$ is the number of lifetime sexual partners for the i-th female, $i = 1, \ldots, F$.

These lists will be used to perform the connections of males and females and build the network. Note that, in Table 2, there are more females than males with few LSP (comparing male and female percentages). It implies that there will be few women with a very large number of LSP. This fact suggests us to start the assignment procedure with women with the largest LSP. Otherwise, it would be possible that, when we have to assign LSP of men to a female with a large number of LSP, there will not be enough men with free sexual partners to be assigned and, for this female, it would be impossible to satisfy the condition that the degree of each node was the number of its LSP.

2.3. Adding Edges to the Network

The procedure in the previous section determines the network as a bipartite graph of females and males, where each node has its predetermined degree (LSP). Now, we have to assign women-men edges in such a way that the total number of edges is as close as possible to the total number of male and female LSP p_m. In other words, preserving as much as possible the predetermined degree of each node as expressed in Eq. (1). Note that this defines an optimization problem in which the optimal value is 0, representing the perfect match between men and women. However, we cannot ensure that such a solution exists (i.e., with a value of 0), and we want to obtain a solution as good as possible in an efficient way. We therefore apply a heuristic method to deal with this combinatorial optimization problem.

First, we are going to take the lists *kFemale* and *AgeFemale* and order them in descending order of the number of LSP. This way, we start assigning first the females with more LSP to free men LSP avoiding the aforementioned inconvenience when females with a lot of LSP are assigned at the end of the procedure when there are not enough free men LSP.

We propose a heuristic assignment process partially based on the Greedy Randomized Adaptive Search Procedure, or simply called GRASP methodology [15]. Specifically, we combine a weighted function with a pseudo-random selection. Note that exact or enumerative procedures, such as Khun-Munkres method [16] cannot be applied here, since their computational cost do not allow the building of medium and large networks. For example, they would need a whole computing day to build an assignment with only 1000 nodes.

We cannot assign women-men randomly if we want to obtain a likely network. Therefore, we are going to make some decisions taking into account that more complex conditions imply more computational time. This is important because our objective is to build very large networks. Our assumptions are:

- People use to join with people with similar habits (principle of psychological similarity [17] or assortativity). Therefore, we are going to define a weight function assuming that: women with few LSP usually match men with few LSP; people with 4 or more LSP use to join with people with 4 or more LSP; and couples where one of them has a lot of LSP and the other few LSP will be uncommon. Then, if x and y are the number of LSP of a woman and a man, respectively, we define the following weight function:

$$\pi(x,y) = \begin{cases} |x-y| & x,y \leq 4 \\ 0 & x,y > 4 \\ 100 & \text{otherwise} \end{cases} \quad (2)$$

- We are not going to consider the age difference in couples. Even though we have data about how Spanish couples join (average difference of 1.5 years with a standard deviation of 5 years) [18], we do not have data and still do not know how these networks evolve over the time, (i.e., breaking edges and creating new ones preserving a likely LSP structure). This can be a good point to extend our model in a future work. Therefore, our priority is the network topology rather than an accurate age distribution. This is because we want to study the spread of STIs diseases, and the diseases spread throughout the edges and the edges define the network structure.

In the procedure to build the women-men assignments we consider the following lists and parameters:

- $kMale$, list with the number of LSP of each male.

- $kFemale$, list with the number of LSP of each female ordered in descending order.

- $\alpha = 0, 1, 2, \ldots$, is a measure of randomness in the women-men assignment, a bigger α implies more freedom in the women-men assignment.

- $matching$, empty list where we are going to store the women-men assignments.

- $krem$, is the list to annotate the number of non-assigned men LSP.

The algorithm to build the network structure by matching men and women can be divided into the following steps:

1. The list $matching$ is initialized as an empty list.

2. Let $krem = kMale$ is initialized with the number of non-assigned men LSP.

3. To proceed with the assignments we take into account every female from $i = 1$ to F (for all females ordered in descending LSP order) and for the female, $i - th$, we consider every male from $j = 1$ to M.

4. If $krem(j) \neq 0$ (male j still has free LSP).

5. For every possible couple (i, j) we calculate the weight of the current edge $\pi_{i,j} = \pi(kFemale(i), kMale(j))$ and add the pair $(j, \pi_{i,j})$ to the W list.

6. Order the W list in ascending order by the value of the weights $\pi_{i,j}$.

7. Take the first $kFemale(i) + \alpha$ pairs of W and select $kFemale(i)$ of them randomly (Notice that if $\alpha = 0$ the technique is greedy, otherwise is GRASP). Increasing the value of α the weight function π constrains less the condition about the number of LSP in the women-men assignment. Let us name W_i the set of the first element of the $kMale(j)$ pairs randomly selected, i. e., we consider only the male nodes component of the pairs, not the weights.

8. Add the edges (i, k) for all $k \in W_i$ to the list $matching$ and update the $krem$ list as follows: $krem(k) = krem(k) - 1$ for all $k \in W_i$.

9. Proceed with the next woman.

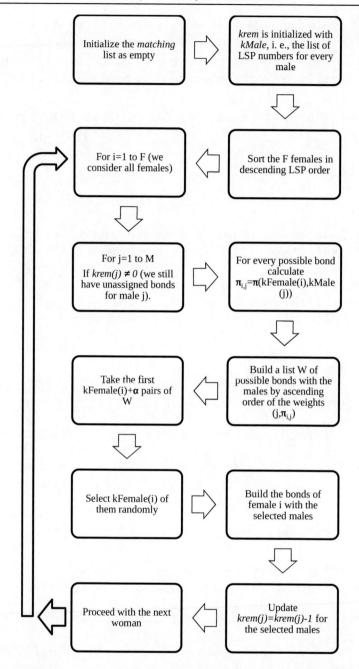

Figure 2. Flow diagram for the algorithm corresponding to the building of the network.

This algorithm is shown graphically in Figure 2.

To sum it up, we assign partners to the males from the still unassigned females, preferentially with those with the lowest values of the matching function $\pi_{i,j}$. Randomness is introduced by allowing for the selection of partners within a set whose size is controlled by the parameter α. The iteration on the males is performed until every individual in the network has been linked. In all our experiments, the edges are assigned satisfying completely

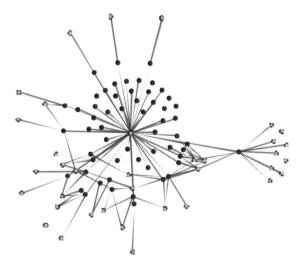

Figure 3. Lifetime sexual network for 100 individuals generated by using the algorithm described in this paper and the statistical data for Spain.

the number of LSP of all the nodes. Results for the networks generated using this procedure are discussed in the next section.

3. Results

We are interested in building a mathematical model for a lifetime sexual partner network representing real statistical data on the number of partners for individuals, as a first step to simulate the evolution of sexually transmitted diseases and, in particular, the human papillomavirus or HPV and the efficacy of vaccination campaigns [19]. The high impact on the population of massive vaccination campaigns against STIs cannot be accounted for by the traditional models based upon differential equations because, in this kind of diseases, it is clear that individuals with many partners are the primary factors in the spreading of the infection.

By using the algorithm described in this paper it is possible to obtain networks encompassing a large number of individuals in a reasonable amount of computation time. In Figure 3 we show an example of a network generated by the algorithm described in the previous section. Notice that, at the core of the network, a women with a high connectivity degree is found, which implies that these networks share some characteristics with the prototypical scale-free networks [20]. We also observe that many individuals in the network are connected with the core of the network through a single relationship. This behaviour has been anticipated in the idea of the core infection model but, to the best of our knowledge, it has not been deduced, in general, from real statistical data. This also means that STIs can also reach people who are into a single relationship in their whole lives because most of the population belongs to the cluster.

We performed an experiment to see the evolution of the CPU time as the network size increases. Thus, we used a computer with CPU Intel®Xeon®X3430 2.40 GHz under MS-

Windows®7 and 16 GB of RAM to build the LSP networks. The code was implemented in C++ and we built 143 LSP networks varying the three parameters as:

- N (the number of people) takes the values $5000, 10,000, 25,000, 50,000, 100,000, 150,000, 250,000, 500,000$ and $1,000,000$.

- $\langle k \rangle_m$ (the average number of sexual partners of men) takes the values 4.5, 5, 5.5 and 6.

- α (the measure of randomness in the women-men assignment) takes the values 0, 10, 20 and 30.

We realized that the CPU time used is very similar for the same N. Then, we decided to compute the mean CPU time for the same N and draw a graphic of the CPU time against the network size N. The graphic can be seen in Figure 4. The complete results can be seen in the web site http://lsp.imm.upv.es.

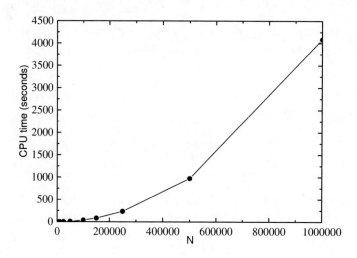

Figure 4. Computation time in seconds vs number of nodes in the LSP network.

It is also interesting to analyze the distribution of the number of sexual partners both for men and women in the network according to the results of the algorithm. In Figure 5 we show the distribution of k, the connectivity degree, for men, women and both sexes. These results correspond to $N = 100,000$ individuals with an average men connectivity $\langle k \rangle_m = 6$ and $\alpha = 10$, i.e., we used a GRASP based technique. The distributions are similar in shape, decreasing algebraically for large k. However, to obtain the exponent of the power-law distribution we will require simulations with a larger number of individuals and this cannot be achieved with the present computational resources. A power-law distribution in sexual networks have also been deduced in other studies [9]. We must also remark that in our technique the ages of the individuals are not an important parameter because we consider lifetime sexual partners and not a snapshot of the network at a given time and the corresponding evolution as relationships are created or destroyed. Nevertheless, it is informative to show the histogram of ages of the partners related by the algorithm in the network. In Figure 6 we show the results. The maximum number of partners are formed by men in the range of 35-40 years old and women from 30 to 55 years old.

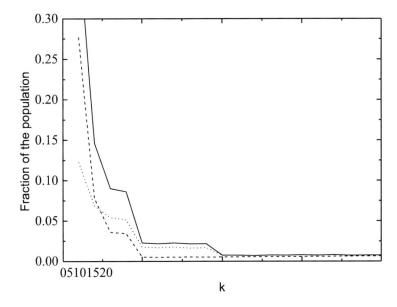

Figure 5. Fraction of individuals with k partners during his/her lifetime for men (dotted line), women (dashed line) and the total population (solid line) as obtained with the algorithm described above. We considered a case with $\langle k \rangle_m = 6$ and $\alpha = 10$.

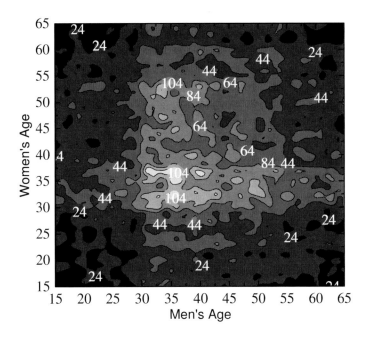

Figure 6. Density map for the couples of sexual partners obtained in the model in terms of the ages of men and women. This plot corresponds to a network with 50,000 individuals and the numbers in white indicate the cases for a given combination of ages.

Supporting Information

Graphs for typical configurations and a table detailing the computational time for generating them are described at the webpage: http://lsp.imm.upv.es. A *Mathematica* program to visualize the networks is also included.

Acknowledgments

The authors gratefully acknowledge J. Díez-Domingo for useful comments and discussions.

References

[1] R. Lozano, M. Naghavi, K. Foreman, S. Lim, K. Shibuya, V. Aboyans V et al., Global and regional mortality from 235 causes of death for 20 age groups in 1990 and 2010: a systematic analysis for the Global Burden of Disease Study 2010, *The Lancet* 380 (2012) 2095-2128.

[2] World Health Organization, Emergence of multi-drug resistant *Neisseria Gonorrhoeae*. Threat of global rise in untreatable sexually transmitted infections. WHO/RHR/11.14, 2011.

[3] D. M. Parkin, The global health burden of infection-associated cancers in the year 2002, *Int. J. Cancer.* 118(12) (2006) 3030-3044.

[4] UNAIDS factsheet, http://www.unaids.org/en/resources/campaigns/globalreport2013/factsheet (Accessed on December, 12, 2014).

[5] A. Burchell, H. Richardson, S. M. Mahmud, H. Trottier, P. P. Tellier, J. Hanley, F. Coutlée, E. L. Franco, Modeling the Sexual Transmissibility of Human Papillomavirus Infection using Stochastic Computer Simulation and Empirical Data from a Cohort Study of Young Women in Montreal, Canada, *American Journal of Epidemology* 169(3) (2006) 534-543.

[6] F. Liljeros, C. R. Edling, L. A. Nunes, H. E. Stanley, Y. Åberg, The web of human sexual contacts, *Nature* 411 (2001) 907-908.

[7] P. S. Bearman, J. Moody, K. Stovel, Chains of Affection: The Structure of Adolescent Romantic and Sexual Networks, *American Journal of Sociology,* 110(1) (2004) 44-91.

[8] S. Helleringer, H. P. Kohler, Sexual network structure and the spread of HIV in Africa: evidence from Likoma Island, Malawi, *AIDS* 21 (2007) 2323-2332.

[9] B. V. Schmid, M. Kretzschmar, Determinants of sexual network structure and their impact on cumulative network measures, *PLOS Computational Biology* 8(4) (2012) e1002470.

[10] Valentian Institute of Statistics, http://www.ive.es.

[11] Encuesta de salud y hábitos sexuales 2003. Instituto Nacional de Estadastica, http://www.ine.es.

[12] A. Chandra, W. D. Mosher, C. Copen, Sexual Behavior, Sexual Attraction, and Sexual Identity in the United States: Data From the 2006-2008. National Survey of Family Growth, National Health Statistic Reports, No. 36, March 3, 2011. http://www.cdc.gov/nchs/data/nhsr/nhsr036.pdf

[13] Key Statistics from the National Survey of Family Growth, http://www.cdc.gov/nchs/nsfg/key_statistics/n.htm.

[14] T. H. Cormen, C. E. Leiserson, R. L. Rivest, C. Stein, *Introduction to Algorithms*, MIT Press and McGraw-Hill, 1990.

[15] T. A. Feo, M. G. C. Resende, Greedy Randomized Adaptive Search Procedures, *Journal of Global Optimization*, 6(2) (1995) 109-133.

[16] J. Munkres, Algorithms for the Assignment and Transportation Problems, *Journal of the Society for Industrial and Applied Mathematics* 5(1) (1957) 32-38.

[17] D. Gentner, A. B. Markman, Structure mapping in analogy and similarity, *American Psychologist* 52(1) (1997) 45-56.

[18] Pau Miret, *La similitud entre los componentes de las parejas jóvenes en España en la primera década del siglo XXI ¿Cada vez más iguales?* Revista de Estudios de Juventud, No. 90, 2010 (Ejemplar dedicado a: Juventud y familia desde una perspectiva comparada europea), p. 225-255, www.injuve.es/sites/default/files/RJ90-16.pdf.

[19] C. K. Fairley, J. S. Hocking, L. C. Gurrin, M. Y, Chen, B. Donovan, C. S. Bradshaw, Rapid decline in presentations of genital warts after the implementation of a national quadrivalent human papillomavirus vaccination programme for young women. *Sexually Transmitted Diseases* 85(7) (2009) 499-502.

[20] A. L. Barabási, R. Albert, Emergence of scaling in random networks, *Science* 286 (5439) (1999) 509-512.

[21] E. H. Elbasha, E. J. Dasbach, R. P. Insinga, Model for assessing human papillomavirus vaccination strategies, *Emerg. Infect. Dis.* 13(1) (2007), pp. 2841. Available from http://www.cdc.gov/ncidod/EID/13/1/28.htm.

ABOUT THE EDITORS

Lucas Jódar Sánchez, PhD
Professor (Full)
Instituto Universitario de Matemática Multidisciplinar,
Universitat Politècnica de València, Spain

Lucas Jódar Sánchez graduated and is a doctor in Mathematical Sciences from the University of Valencia in 1978 and 1982, respectively. Full University Professor in Applied Mathematics since 1991. He was the head of the Department of Applied Mathematics of the Universitat Politècnica de València from 1992-2002. He is currently head of the University Institute of Research in Cross-disciplinary Mathematics since its foundation in 2004. Dr. Jódar holds five six-year periods of research and has directed 30 doctoral theses. He is co-author of 410 research articles. He has been in charge of competitive research projects since 1988. He is currently in charge of the Spanish team within the European Project entitled STRIKE which is based on the use of computational methods in finance. He is editor in several international journals specialised in topics such as Modelling and Numerical Methods.

Elena de la Poza Plaza, PhD
Assistant Professor
Departamento de Economía y Ciencias Sociales,
Universtitat Politècnica de València, Spain

Elena de la Poza Plaza is an Assistant Professor of Economics at the Universitat Politècnica de València (UPV).

She holds a PhD in Assets Valuation from the UPV. During and after receiving her PhD degree, she spent research periods at the University of North Carolina, University of Tennessee, University of Padova and University of Panama. She has taught courses at Washington State University, but also research workshops at Technical University of Ostrava (Czech Republic) and Faculty of Economic Sciences of Warsaw University of Life Sciences, (Poland).

Dr. de la Poza Plaza has participated as a speaker at over 30 international conferences. Also, she has served as a reviewer for international scientific journals. Her publications are focused on modeling individuals and organizations behavior, but also on economic valuation of assets of investment such as real estate, fine arts and healthcare.

Luis Acedo Rodriguez, PhD
Research Assistant
Instituto Universitario de Matemática Multidisciplinar,
Universitat Politècnica de València, Spain

Luis Acedo Rodriguez is a Researcher at the Institute for Multidisciplinary Mathematics of the Universitat Politècnica de València in Valencia, Spain.

He received his graduate degree and his PhD in fundamental physics at the University of Extremadura in Badajoz, Spain. He has been visiting scientist at the Universities of Utrecht and Mexico. From 1998 to 2006 he was teaching assistant in Physics and Mathematics at the Universities of Extremadura and Salamanca. His current research interests are interdisciplinary: including mathematical modelling of infectious diseases, epidemiology, networks and gravitational physics. He has published more than 70 refereed journal papers and book chapters and he has been co-organizer of ten international conferences on mathematical modelling as well as being the guest editor of several special issues by internationally-recognized publishers. Nowadays Dr. Acedo is participating on a research project for the modelization of meningitis propagation and control and a research contract with the pharmaceutical industry.

INDEX

A

actuarial, 195, 196, 207, 208, 209
adoption, v, vii, 65, 66, 67, 68, 69, 70, 71, 72, 73, 74, 75, 76, 77, 78, 79, 80, 81, 82, 83, 137
aftermarket, 179, 180, 181, 184
aftermarket or aftermath price, 181
aftermath, 179, 181, 183, 184
architecture, vii, 20, 125, 126, 127, 128, 131, 132
attributes, 12, 29, 49, 50, 51, 52, 53, 160, 222
attributes assessment, 49
autocracy, 49, 51

B

biorthogonal wavelet, 206
bowl phenomenon, 221

C

causes, 1, 2, 10, 134, 143, 163, 165, 199, 231, 248
chemists, 86, 90, 91, 92, 93, 94
Colombia, 49, 52, 54, 55, 211
combinable goods, 180
competitiveness, vi, vii, 26, 133, 134, 135, 136, 137, 138, 139, 140, 141, 147, 148, 149, 150, 154, 156, 157, 158, 159, 160, 161, 162, 163
construction, 27, 77, 103, 104, 125, 126, 127, 128, 129, 131, 239
consult, 49, 51
correspondence analysis, 26, 29
Counter-Reformation, vi, 125, 131
Cronbach Alpha, 53
culture of organization, 99, 109

D

Data Envelopment Analysis (DEA), 133, 134, 136, 141, 142, 144, 146, 147, 161, 162
delegation, 49, 51
demographic, 85, 135, 136, 159, 160, 195, 196
diffusion, vii, 65, 66, 67, 68, 69, 70, 71, 72, 73, 74, 75, 76, 77, 78, 79, 80, 81, 82, 83, 167

E

economic and institutional factors, 65, 81
eletronic prescribing system, 86
European Communities, 146
European Union (EU), vi, vii, 133, 134, 135, 140, 142, 146, 147, 148, 149, 150, 156, 157, 159, 160, 161, 162, 163

G

graduation, 195, 196, 197, 199, 202, 203, 204, 205, 206, 207, 208, 209

H

health technology, 65, 66, 67, 68, 69, 72, 75, 76, 79, 80, 82, 83
Heligman, 195, 197, 198, 203, 204, 205, 206, 208
Heligman and Pollard, 195, 197, 198, 203, 204, 205, 206
higher education, vii, 10, 25, 27, 35, 36, 37, 141, 150, 159
human behaviours, 99
human capital, vi, vii, 133, 134, 136, 137, 140, 141, 146, 150, 156, 157, 158, 160, 161, 162

human resources management, 99, 111, 138

I

INCODE questionnaire, 26, 34
innovation competency, 25, 26, 27, 34, 35

J

joint decision, 49, 51

L

leaders, 5, 49, 50, 51, 52, 54, 73, 77, 78, 81, 106
legitimacy, 49, 50, 51, 55
life tables, 195, 196, 203

M

machine variability, 221, 222, 223, 230, 232
Malmquist Productivity Index (MPI), 134, 136, 143, 145, 146, 147, 149, 151, 152, 153, 156
Manizales, vii, 49, 52, 54, 55
manufacturing systems, 221
membership fee, 177, 178, 179, 182, 183, 184
metabolic syndrome, vii, 57, 58, 63
m-health, 37
m-learning, 37, 38, 39, 40, 41, 42, 43, 44, 45
mobile devices, 37, 38, 41, 46, 47
mortality rates, 195, 196, 197, 203, 207, 208
moving objects, 211, 218
multi-homing, 177, 178, 179, 180, 181, 182, 183, 184
multiple demand, 177
multiresolution analysis, 200, 201, 202
multivariate analysis, 26

N

nonparametric graduation, 196

O

organizational values, 99, 111

P

participative leadership, 49, 50, 51
personal behavior, 125

pharmacy, vii, 57, 59, 95
Pollard, 195, 197, 198, 203, 204, 205, 206, 208
polynomial interpolation, 196, 203, 208
power, 2, 4, 49, 50, 51, 54, 55, 112, 128, 138, 142, 197, 236, 246
prevention of chronic diseases, 57
price per service, 177, 178, 183
public university system, vii, 1, 22

Q

quality, vii, 4, 45, 57, 60, 63, 66, 75, 76, 78, 80, 86, 93, 94, 105, 106, 107, 108, 110, 134, 137, 140, 141, 147, 150, 156, 157, 158, 159, 160, 178, 187

R

recommendations, vii, 1, 2, 28, 58, 62
Regional Competitiveness Index (RCI), 133, 134, 136, 140, 141, 142, 146, 147, 148, 149, 156, 157, 159, 161
reliability, 40, 49, 53, 54, 106, 172, 191
risk of death, vi, 196, 197, 199, 204

S

satisfaction, vii, 3, 85, 86, 87, 89, 92, 93, 94, 101, 105, 106, 108, 112, 179
semantic trajectories, 211, 214, 215
shaping the culture, 99, 109
smoothness, 197, 205
social factors, 59, 65, 81
stakeholders, vii, 49, 50, 51, 52, 53, 54, 77
students, vii, 2, 3, 4, 6, 7, 9, 10, 12, 13, 14, 20, 22, 25, 26, 27, 28, 34, 35, 36, 37, 38, 41, 42, 43, 45, 46, 47, 48, 51, 54, 128, 158
subscription fee, 178, 179, 180, 183
substitute goods, 177, 179

T

technology nature, 65, 72, 79, 80, 81, 82
threshold, 13, 150, 161, 185, 186, 190, 202, 205, 212, 217, 232
thresholding, 196, 202, 206
throughput line, 221
trajectory similarity, 211, 212, 218, 219
two-part tariff, 177, 178, 180, 184

U

University stakeholders, vii
urgency, 49, 50, 51, 53, 55
user, 22, 36, 85, 86, 87, 89, 93, 179, 182, 217, 232

V

Valencia, vi, vii, viii, 1, 25, 49, 60, 85, 96, 97, 113, 114, 115, 116, 121, 125, 126, 127, 128, 129, 130, 131, 132, 165, 195, 221, 228, 232, 235, 237, 238, 239, 251, 252
Valencian Community, vi, 10, 85, 86

W

wavelet graduation, 195, 196, 197, 199, 202, 203
wavelet(s), vi, viii, 195, 196, 197, 199, 200, 202, 203, 204, 206, 207, 208, 209
wrong behaviour, vii, 1